COBB VALLEY:

A GEOLOGICAL GUIDE

R.A. Cooper

Guidebook No. 6

ISBN 0-908678-03-7

Published by the Geological Society of New Zealand
C/o PO Box 30-368
Lower Hutt
New Zealand

Printed by Deslandes Ltd, Petone

Fig. 2 Aerial view down Cobb Valley from Lake Cobb. The sandstone beds in the foreground are discussed on p. 13. Photo: DLH.

ABOUT THIS GUIDEBOOK

Visitors to the Cobb Valley will be impressed by the deep open U-shaped valley form and the even, gently-sloping surface (peneplain) forming the mountain tops from Mount Peel down to the Mount Arthur Tableland. The U-shaped valley form was carved during the Ice Age, and the peneplain surface was produced by extensive erosion about 60 million years ago. The rocks that make up the mountains themselves are very much older and contain New Zealand's oldest fossils—an extinct group of slater-like marine animals called trilobites. In the first chapter I discuss why the Cobb Valley is of such geological importance; the second, third, and fourth chapters discuss the geology of the Cobb Valley in detail; and the final chapter is a trampers' guide, pointing out features of interest and describing seven walks through the valley area. The glossary at the back will help explain unfamiliar terms.

My hope is that, after reading this book, more people will appreciate the Cobb Valley for itself and understand its interest and importance to geologists, amateur naturalists, and indeed for all who enjoy the outdoor life.

CHAPTER 1—INTRODUCTION

WHAT IS SO SPECIAL ABOUT COBB VALLEY?

Cobb Valley is one of the few places accessible by road where New Zealand's oldest rock successions can be examined. Much of the earliest geological history of this country is recorded in rocks which outcrop in the valley—the country's oldest fossils are found here, as is evidence of the oldest volcanic activity in the country, including ancient lava flows. The comparatively small area contains a great variety of rock types, themselves reflected in a variety of soil types and in the varied plant species associated with these soils.

A wide variety of unusual minerals and rocks is found here, including deposits of asbestos, magnesite, and chert, and the valley provides many excellent examples of the effects of glaciation in a mountainous region. Finally, the topography of the Cobb—Mount Arthur Tableland is especially interesting to the geomorphologist (who studies the origin of land forms), because it shows evidence that the whole Northwest Nelson region once formed part of a continuous lowland plain.

The geological complexity of the area is extraordinary and indicates the wide range of environments which have followed each other over the years. Changes in climate, in sea level, and in the position of rocks relative to each other can to a great extent be documented and dated by geologists today.

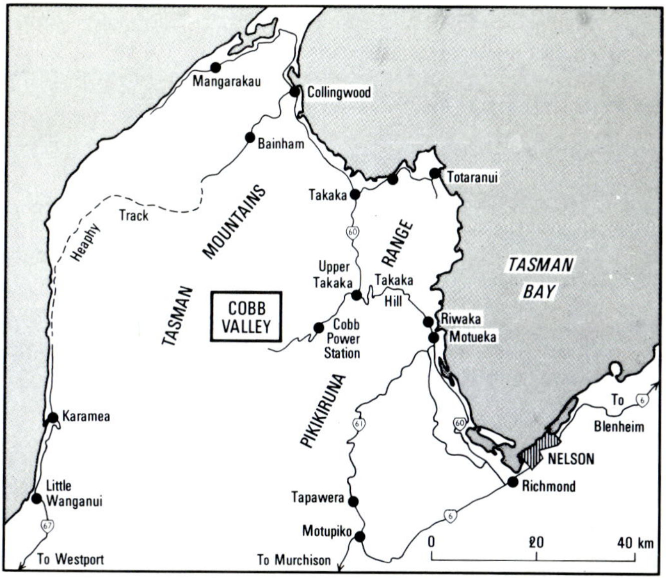

Fig 3 How to get there. From the turn-off at Upper Takaka on State Highway 60, a Ministry of Energy Electricity Division road extends for 28 km to Cobb Valley. The road is narrow but is no problem if you drive carefully. It is not recommended for caravans.

GEOLOGICAL TIME—THE GEOLOGIST'S PERSPECTIVE

The geologist must learn to appreciate the immensity of time in the history of the earth. Most of us can envisage a 100-year time span (back to the eruption of Mt Tarawera, for example) or even one of 1000 years (to William the Conqueror). 10,000 years takes us to the earliest-known civilisation; 20,000 to the last ice age during which the Cobb Valley was filled by a glacier; 1,000,000 years, more than half-way back to the beginning of the present ice age.

If 1,000,000 years is an immensely long time to most of us, consider the age of rocks forming Cobb Valley, which ranges from 450 to 500 million years. By this time, the exact number of years is impossible to determine and therefore rather meaningless in ordinary speech. Figure 5 illustrates the huge time span covered by geological history, and Figure 7 gives the names generally used for subdivisions of the later part of the span—the geological time scale.

There are two main methods of "dating" rocks. The first and most widely used method relies on the fossils contained by the rocks. Because plant and animal species change with time, a process we call evolution, each geological stage has some characteristic species found in no other stage. Further, some of the species were able to migrate across wide regions and even around the world. Thus, the sequence of fossil graptolite and trilobite species (discussed in the next chapter) in New Zealand is very similar to that found in Australia, which enables us to correlate our rocks with Australia. By this process, fossiliferous rock sequences around the world are correlated with each other, including the "standard" rock sequences (mainly in Britain and Europe) whose names are shown in Figure 7—Cambrian, Ordovician, Silurian, etc. This dating method enables us to place our Cobb Valley rocks within a standardised international rock sequence (the geological time scale), but it does not tell us how many years old they are. For that we must use the second dating method.

Fig. 4. Lake Sylvester occupies a cirque (explained on p. 31) just above the bushline. Ice-rounded outcrops of Lockett Conglomerate (p. 11) in foreground. View looks toward Pikikiruna Range. Photo: LL.

GEOLOGICAL TIME SCALE

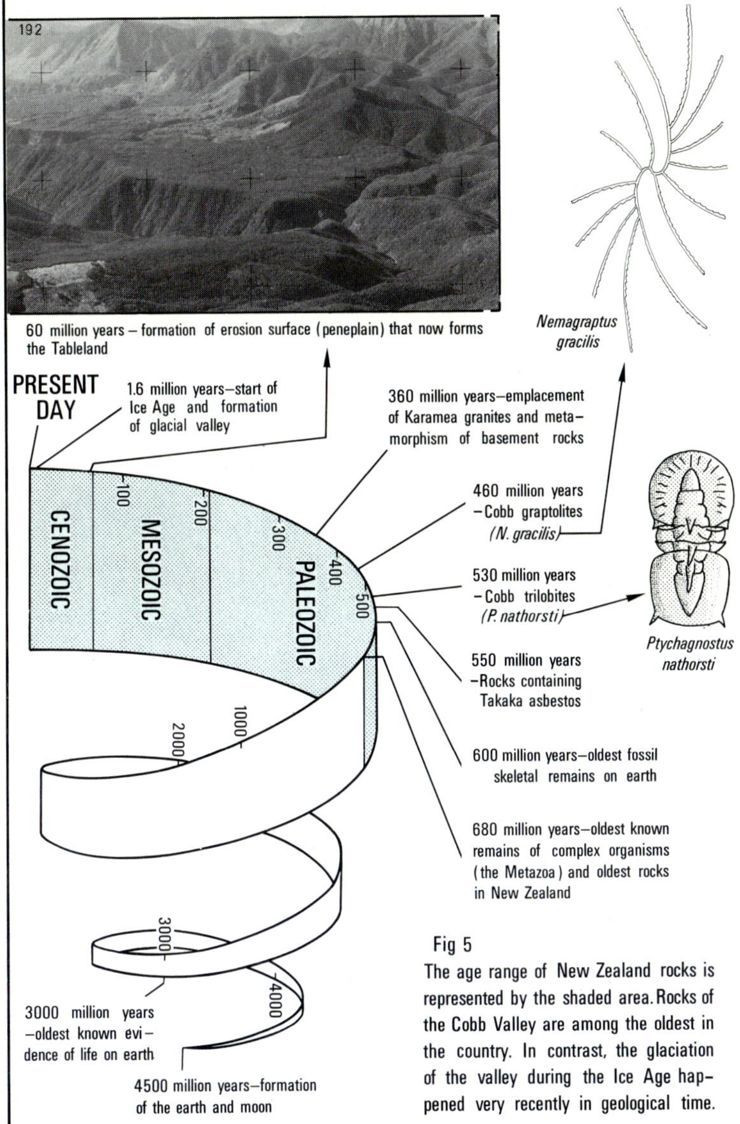

Fig 5

The age range of New Zealand rocks is represented by the shaded area. Rocks of the Cobb Valley are among the oldest in the country. In contrast, the glaciation of the valley during the Ice Age happened very recently in geological time.

The second method uses the minute amounts of radioactive isotopes contained in the minerals of some rocks. Isotopes are peculiar variants of such normal elements as lead, argon, carbon, and strontium, many of which are produced by radioactive decay of other elements. For example, normal lead has several isotopes, including ^{206}Pb which is formed by the radioactive decay of uranium (^{238}U). By measuring the proportion of 'parent' (^{238}U) to 'daughter' (^{206}Pb) in the rock, we can calculate its age in years. If every rock could be dated this way there would be no need to use fossils for dating. However, most rocks cannot be dated by this method; some do not contain the right isotopes, and others have been hopelessly altered by heat or chemical changes. There are, however, enough data to roughly calibrate the geological time scale, and the age in millions of years is shown in Figure 7.

For rocks older than the oldest fossils there is no other reliable means of estimating age. The method was also used to date rocks recovered from the moon; their age was found to be 4500 million years, agreeing with previous estimates of the age of the Earth.

Fig. 6 View across Cobb Valley from Iron Hill ridge on line with the fault separating scree-covered slopes of Mount Peel (Tasman Formation) and craggy slopes of Mount Mytton (Summit Limestone). Photo: LL.

THE STORY OF COBB VALLEY

Because it has had a very complex geological history, it has taken us a long time to piece together the sequence of events that has moulded the valley and its rocks into their present form. Even now there are many unanswered questions, and much work is still in progress. Some detailed studies of the area are listed under *Further reading,* and a major source of information on the geology of Northwest Nelson is the 1:63 360 (1 inch = 1 mile) geological map series published by the New Zealand Geological Survey.

The most obvious features of the valley are those most recently formed. The deep and smooth-sided valley form was produced during the Ice Age (Pleistocene Period, 0-1.6 million years ago, see p. 30). In an earlier chapter of Cobb Valley history, the tableland surface was formed as part of a lowland plain during a long period of erosion (60 million years ago, see p. 27).

From here, however, it's a long way back to the next earlier chapter—the formation, folding, uplift, and erosion of the far older, Early Paleozoic rocks (350-550 million years) which form the valley's foundation, or "basement" (p. 8).

We know, from Parapara Peak in Golden Bay, that there was yet another chapter: submergence of the land, formation of sedimentary rocks, folding and uplift of the countryside in late Paleozoic time (250 million years ago). But although these events must have affected the Cobb area, none of the rocks are found preserved in Cobb Valley today and we will not discuss this chapter in detail. It is worth noting that only a small fragment of this late Paleozoic sedimentary sequence is now preserved at Parapara Peak, the rest having been removed by erosion. If this fragment, too, had been lost, geologists would know nothing of the submergence of the land and formation of sedimentary rocks at this time—a sobering reflection.

THE BALANCE OF UPLIFT AND EROSION

Any land area is the net result of two opposite processes: uplift of the land (or lowering of sea level) and erosion of the land surface. The two go together. Without uplift there can be no erosion. When uplift is rapid and continues for a long time, mountains are produced and erosion becomes intense, as in the Southern Alps. If uplift stops and erosion continues, the end result is a lowland plain such as that drained by the Murray and Darling rivers of Australia. If uplift reverses, that is, if the land starts going down beneath the sea, the result is called a drowned topography, one example of which is the Marlborough Sounds with its drowned river valleys.

The Cobb Valley area demonstrates both uplift and erosion. The mountains have been formed by uplift in the last 10 million years or so (see Chapter 3), and erosion by chemical (soil weathering) and mechanical means (freeze and thaw, glacial scouring, river abrasion) has produced a vast amount of sediment that has been carried down the Cobb River to the Takaka, where much of it was deposited below Upper Takaka as huge gravel fans. Subsequently the Takaka River has cut into the fans, and most of the material has been carried down to the coast. Remnants of the fans can be clearly seen on both sides of the Takaka Valley from State Highway 60 at Takaka Hill.

CHAPTER 2—THE PALEOZOIC ROCKS

THE THREE BELTS

New Zealand's oldest rocks are found in West Nelson, Westland, and Fiordland. The Fiordland region has been offset by 450 km from West Nelson by the Alpine Fault (Fig. 8), a major fault line which goes through the entire thickness of the Earth's crust, and which can be traced from the Macquarie Ridge (in the sea south of Fiordland) to the Kermadec Trench (north of East Cape). Several of the distinctive West Nelson Paleozoic rock formations discussed below have recently been discovered in Dusky Sound; more links are likely to be found as the geological exploration of Fiordland continues.

Apart from rocks in some small areas at Charleston and in the Victoria Range, 100 km south of Cobb Valley (see p. 21), the oldest rocks of the country are of Cambrian age. They lie in a narrow belt of country extending from the Wakamarama Range south to the Wangapeka Valley, passing right through Cobb Valley (Fig. 8). This belt, called the Central Sedimentary Belt (Central Belt, for short), contains abundant volcanic rocks and limestone; it contrasts with the Western Belt, which is represented in the upper part of Cobb Valley, upstream from Chaffey Hut, and which contains no volcanics and no limestone. A major, ancient fault, the Anatoki Thrust, separates the two belts (Fig. 17). It is not yet clear how the two belts ended up adjacent to each other, but it is evident that their sediments formed in distinctly different environments, not originally close together. The original position of a third belt, the Eastern Belt, is also unclear, but at present it is possible that it was originally part of the Central Belt sequence. The Eastern Belt includes the huge limestone and marble bodies forming the Pikikiruna Range, Mount Arthur, and Mount Owen.

In Cobb Valley, rocks of the Central and Western belts are particularly well displayed. The geology west of the Anatoki Thrust (Western Belt) is completely different in sediment type and structure from that east of the thrust (Central Belt). The Central Belt sediments formed along a volcanic chain lying some distance offshore in Cambrian time, whereas the Western Belt sediments formed in a basin or slope adjacent to an eroding continental region. From reconstructions of the Earth's continents in early Paleozoic time (Fig. 10), we can be reasonably sure that this continental region was that which today forms part of Southeast Australia and Antarctica. So, the New Zealand Western Belt rocks are now widely separated from their source area.

THE ROCK SEQUENCE IN COBB VALLEY

The various rock formations and a brief description of each are given in Table 1 (inside back cover). Within the Central Belt sequence the oldest rocks are mapped as the *Balloon Formation* (named after Balloon Creek).

DEPOSITION AND EROSION

Repeated changes in the pattern of land and sea

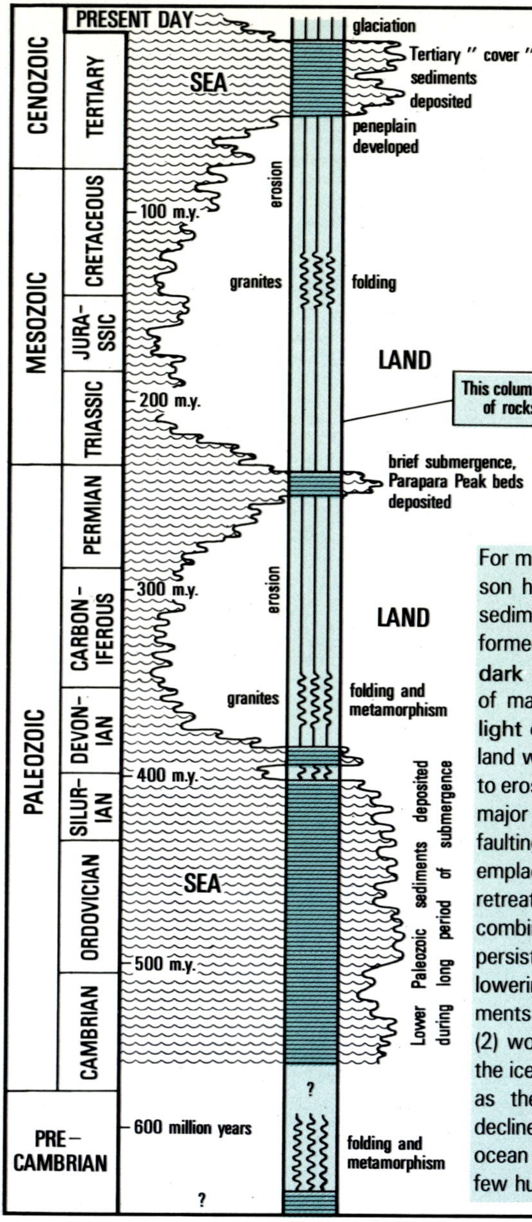

For much of its history, Northwest Nelson has been under sea; most of its sedimentary rocks are marine - that is, formed under the sea. In the column the **dark colour** shows the age range of marine sedimentary rocks and the **light colour** show the times when the land was above sea level and exposed to erosion. Zig-zag lines show when the major episodes of deformation (folding, faulting), metamorphism and granite emplacement occurred. Advance and retreat of the sea was caused by a combination of two factors: (1) gradual, persistent earth movements raising and lowering the whole land surface (movements of up to thousands of metres); (2) worldwide changes of sea level as the ice caps expanded and retracted or as the mid-ocean ridges grew and declined, changing the volume of the ocean basins (movements of up to a few hundred metres).

Fig. 7 Although Northwest Nelson is only a small part of New Zealand, the time represented by its rocks is equal to about one-third of New Zealand's known geological history.

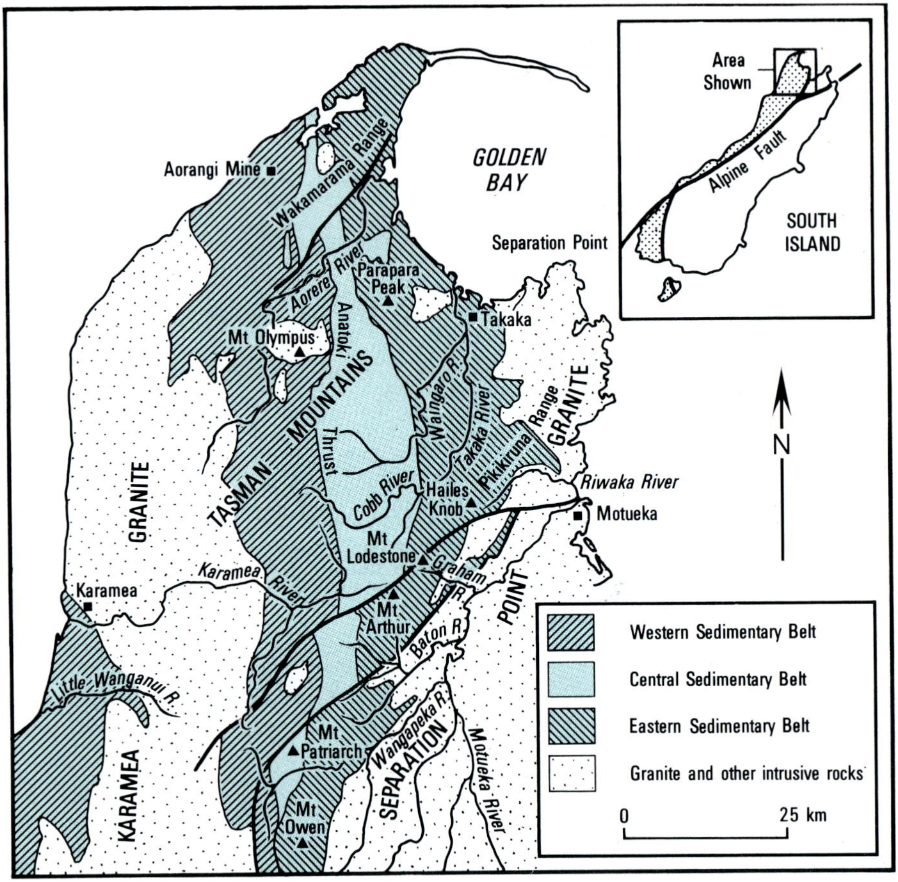

Fig. 8 The main Paleozoic rock belts of Northwest Nelson. Inset shows distribution of lower Paleozoic rocks in the South Island.

Fig. 9 Bluff formed of Lockett Conglomerate in the cirque wall above Iron Lake. Two intersecting sets of cleavage planes cause the rock to break into angular blocks (see p. 11). Bedding dips gently to left. Photo: LL.

They outcrop along the shores of the Cobb Reservoir (see map, Fig. 19) and are of Early or Middle Cambrian age. They are a mixture of dark siltstone, sandstone, and chert—generally well bedded but in some areas fragmented and jumbled up to form a type of conglomerate called diamictite. Good outcrops of diamictite with angular sandstone fragments (orange when wet) can be seen by the lake edge near the head of the reservoir. Chert is a splintery, glassy rock composed entirely of silica; it is found in great quantity in the disused chert quarry on top of the ridge 2 km NE of Cobb Dam. A band of volcanic rock outcrops at the head of the reservoir (locality 7, Fig. 30).

The Balloon Formation was followed by a major volcanic episode, represented by the rocks of the *Devil River Volcanics* formation. These are well exposed in the cirque wall above Lake Peel and in Cobb River between Myttons Creek and Chaffey Stream, where ancient submarine lava flows are found. The volcanic rocks are commonly green because of the presence of the mineral chlorite, which formed during the mild metamorphism of the entire Paleozoic sequence. Grey laminated siltstone of the *Tasman Formation* was the next to be deposited, with limestone lenses containing late Middle Cambrian trilobites (see p. 13). Tasman siltstone forms ice-smoothed hummocks in the valley floor at the Trilobite Creek camping ground; the richest fossiliferous trilobite limestone lens is that of Trilobite Rock.

The overlying *Lockett Conglomerate* forms the ridge top at Iron Hill and is a most distinctive and attractive rock (Fig.9). Rounded pebbles up to about 15 cm in diameter, of chert and old volcanic rocks are clearly seen; they represent the deposits of an ancient gravel fan. The *Anatoki Formation,* mainly sandstone and siltstone, is next in the sequence and is best seen in Diamond Lake Stream. However, the overlying *Patriarch Formation* is barely represented at all in Cobb Valley; it is best developed in the ridge north of Mount Patriarch near Wangapeka Valley.

A change to limestone deposition took place just after the close of Cambrian time and lasted until near the end of the Middle Ordovician. The *Summit Limestone,* deposited during this time, forms the impressive bluffs on Mount Mytton; it is also seen as a discontinuous chain of lenses along the Anatoki Thrust. One such lens stands up as a huge blade, known to geologists as 'The Gendarme', on Lockett Range in the saddle northeast of Mount Benson (Fig. 17). The limestone contains tiny microfossils called 'conodonts' (Fig. 14), which are used to date it. (They are extracted by dissolving the limestone in acid, which leaves the conodonts in the insoluble residue.) The youngest rocks in the Central Belt also appear in the Eastern Belt; they are mainly dark siltstone and quartz-rich sandstone—*Wangapeka Formation.* They contain poorly-preserved graptolites in the Peel Range and are of Upper Ordovician age.

The Western Belt sequence is much less varied in rock type and consists of a rather monotonous succession of alternating beds of quartz-rich standstone and grey siltstone. The four formations mapped—*Roaring Lion, Leslie, Douglas,* and *Peel*—appear as north-south-trending strips on

New Zealand in Cambrian Time

Fig 10

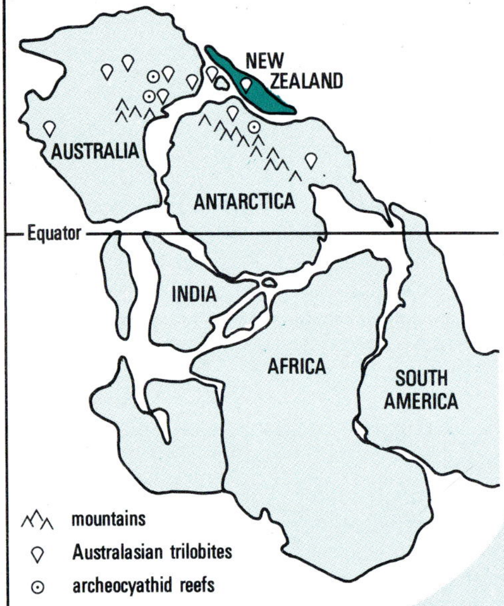

In Late Cambrian time (520 million years ago) the southern continents together formed the supercontinent Gondwanaland. New Zealand lay along the margin of Gondwanaland but of course, at that time, did not include the younger rocks which make up most of the country today. Its shape is thus unrecognisable in the figure. Antarctica (minus its ice sheet) lay on the Equator and New Zealand and Australia both lay in tropical latitudes in the Northern Hemisphere. It is not surprising, therefore, that warm water reef-building fossils (archeocyathids) are found in both Antarctic and Australian rocks. In Middle Cambrian time, shallow seas covered large areas in Australia and Antarctica; New Zealand formed a belt of volcanic offshore islands. Limestones were formed in all three areas and they contain a distinctive fauna of trilobites one of which, *Koptura*, is figured. Others are shown in the section on trilobites. At the end of the Cambrian Period a major episode of mountain building was under way in South Australia and Antarctica. Quartz-rich sands and silts were brought down to the sea by rivers draining these mountains, and deposited over southeastern Australia, New Zealand (Western Belt) and parts of Antarctica.

The tail of a Middle Cambrian trilobite, *Koptura*, found in Australia, New Zealand and Antarctica.

The world at the close of Cambrian time. The supercontinent, Gondwanaland, straddles the Equator.

the map. The *Leslie Formation* contains black slate, which is rich in graptolites at several localities, especially in the ridge south of Waingaro Peak above Fenella Hut, and on the ridge south of Chaffey Stream (see p. 45). Near the base of the Leslie Formation is a thick quartzite unit that can be traced northwards to Mount Olympus and south to the Wangapeka Valley, generally as two distinct bands (Fig. 35). The *Roaring Lion Formation* is composed of alternating sandstone and siltstone beds, which commonly form rugged topography on ridge crests and parallel ridges as at Island Lake (best seen from Mount Gibbs; see also Fig. 2).

Fig. 11 Band of well-bedded Cambrian limestone (Tasman Formation) in saddle on ridge crest, 2.3 km SE of Mount Benson. Photo: J. Simes.

COBB TRILOBITES

The trilobites of the Cobb Valley are New Zealand's oldest fossils (530 million years). In fact, trilobites are the earliest of all known arthropods—that phylum which includes insects, crayfish, and spiders—and appear in the fossil record near the beginning of the Cambrian, just after the first appearance of fossilised hard parts of organisms. They endured for 350 million years—that is, for about two-thirds of the time from the Cambrian to the present day—but were most diverse and abundant in Late Cambrian time. All trilobites were marine; they are most abundant in rocks of those regions that lay in the tropical zone in early Paleozoic time (Fig. 10). They generally lived by foraging along the shallow sea floor, sifting fine particles of organic matter. They range in size from a few millimetres up to more than 300 mm; most of those found in Cobb Valley rocks are incomplete and range from 3 mm up to 20 mm. The trilobite limestone of Trilobite Rock (on the roadside near the head of Cobb Reservoir) represents an accumulation of carbonate detritus and fossils that have been concentrated by current activity in shallow Cambrian sea waters. Freshly-broken rock surfaces *do not* show the trilobites as well as do broken surfaces that have weathered slightly. Please minimise further damage to the rock and don't hit it with hammers. It is unique in New Zealand.

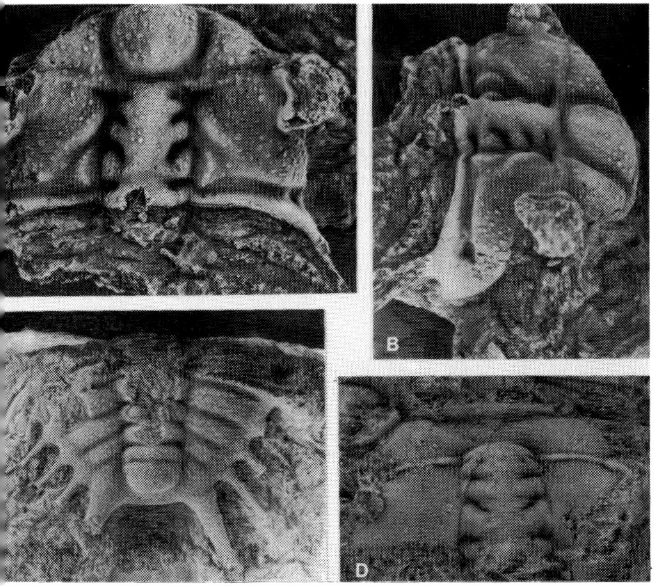

Fig. 12 Trilobites from Cobb Valley. A, B: The head of Nepea; *C, tail of* Dorypyge, *with several spines broken off; D, head of* Pianapsis. *All approx. x3. Photo: RAC.*

The dorsal, or upper, surface of a trilobite was hardened into rigid plates by the mineral calcite and is the only part preserved as fossils. It has two furrows running full length and dividing the dorsal surface into three 'lobes' (hence, tri-lobite). Trilobites had a head, or cephalon, usually with prominent eyes, a thorax which consisted of several segments that were hinged, allowing the body to curl and enroll, and a tail, or pygidium. The division into head, thorax, and tail cuts across the three longitudinal lobes (Fig. 15). Like all arthropods, trilobites had an external skeleton which had to be shed periodically to allow the body to grow in size. A new exoskelton hardened after a short interval and remained until this, too, had to be shed. The tail and thoracic segments were shed as single pieces, but the head region of the skeleton was shed only after it had been split into several fragments. The splits were not haphazard but took place along pre-existing lines of weakness, or sutures. As many as five separate parts were separately shed from the head region. It is these parts together with the tail and thoracic segments that we find preserved in Trilobite Rock today. A complete specimen, with all parts intact, has never been found there.

The trick is to be able to recognise the parts and, particularly, which parts belong to which species. There are parts of at least 16 different trilobites known in the limestone of Trilobite Rock; some of these are illustrated in Figures 12, 13 and 15. Of the 16 species, eight differ from the general trilobite plan as shown in Figure 15, and instead have only two thoracic segments, and a head and tail of almost identical size. They are called agnostid trilobites, and specimens are illustrated in Figures 13 and 15. Agnostid trilobites were generally smaller than the standard version and had no eyes. They are particularly important in Cambrian rocks for correlation and fossil dating; some were planktonic—i.e., they lived in the surface waters of the ocean.

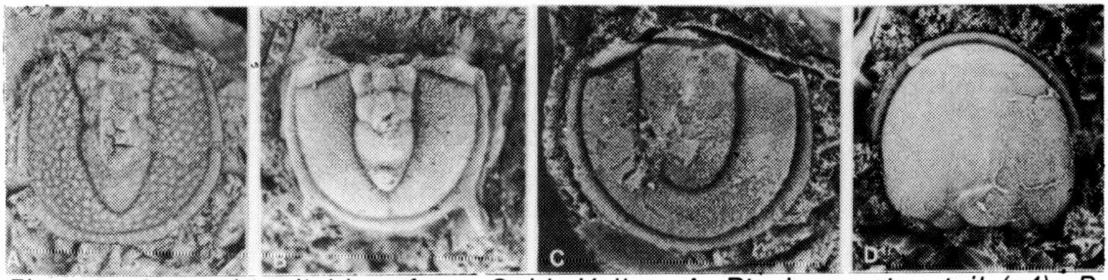

Fig. 13 Agnostid trilobites from Cobb Valley. A, Ptychagnostus *tail (x4); B,* Goniagnostus *tail (x7); C,* Hypagnostus *tail (x7); D, unidentified head (x7). Photo: RAC.*

TRILOBITE EYES

Scientists have long been intrigued by the eyes of trilobites. They are the most ancient visual system known and indeed are the earliest of all well-developed sensory systems. Trilobite eyes were compound and, like the lateral eyes of modern crustaceans and insects, they are composed of radially arranged visual units, fanned out to encompass a wide-angled visual field. The structure and evolution of the eye and its component lenses have received much detailed study. However, there is still considerable debate about just how the eye functioned. Because it was made of calcite, the lens surface was fossilised along with the cephalic skeleton, and we can determine the number and arrangement of individual lenses; but the arrangement of fibres, photosensitive pigments and sensory nerves was not preserved, and we can only guess by comparison with living arthropods. By studying their lens design, however, we can make inferences about the optical system used by trilobites.

Through the long history of trilobites, evolutionary changes in the eye and, particularly, in the lens structure, can be traced. By Ordovician time an eye structure known as aggregate or schizochroal had evolved, in which the individual lenses have a peculiar internal structure. It has only recently been realised that the peculiar structure is a correction for spherical aberration and that the lens design is in fact identical to that proposed (quite independently of any knowledge of trilobites) by Descartes and Huygens. Trilobites had thus found a solution to a major problem with optical systems and so achieved a unique feat in the history of life.

Fig. 14 Other fossils found in the trilobite limestone of Cobb Valley are all very small and most have been found only after the rock has been dissolved away by acid. Tiny black brachiopods are relatively common; they are bivalved like the larger brachiopods, but their shells are composed of chitino-phosphate rather than calcite. They provide useful information on the age of the rocks containing them. Shown above are Acrothele *(A) and* Micromitra *(D). Tiny molluscs, ancestors of today's shellfish, are shown in B, a new genus and species, and F,* Melopegma. *Also shown above, but not from Trilobite Rock, are the tooth-like conodonts* Juanognathus *(C) and* Cordylodus *(E). 20x–40x. SEM photos: John Simes.*

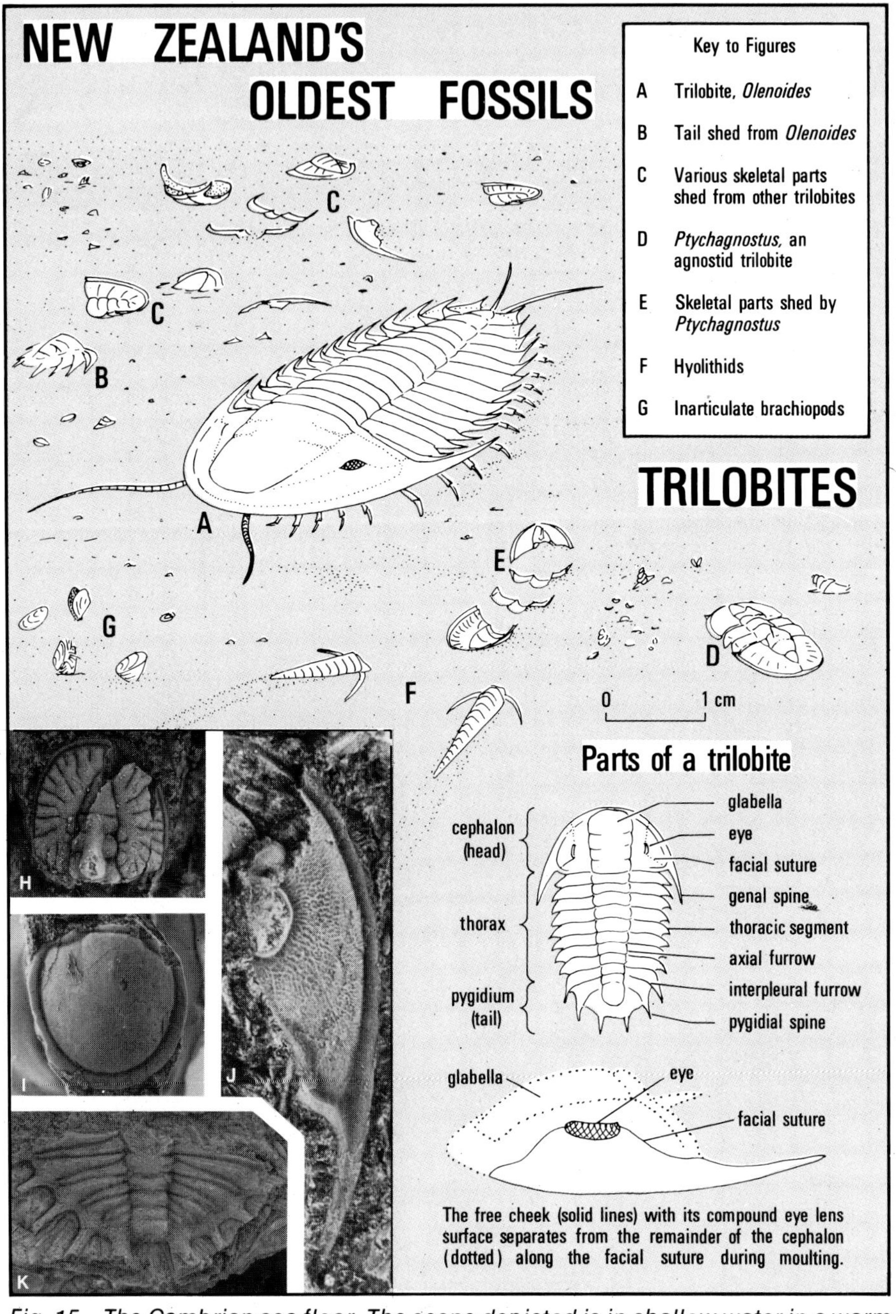

Fig. 15 The Cambrian sea floor. The scene depicted is in shallow water in a warm sea. Trilobites and hyolithids (conical shells) are shown foraging, and the sea floor is littered with shed skeletal fragments of trilobites. The living animal which most closely resembles a trilobite is the horseshoe crab Limulus. *Cobb trilobites in photograph at left are H,* Goniagnostus *(x10); I.* Grandagnostus *(x5); J, free cheek of a papyriaspid (x10); K, tail of* Olenoides *(x2).*

COBB GRAPTOLITES

Like trilobites, graptolites have long been extinct—since the end of the Paleozoic period, 230 million years ago. Unlike trilobites, each graptolite was not an individual animal but a colony of polyps, each one connected with others of the colony by its stalk. Each polyp—or graptozooid—had a pair of tentacle-like lopophores by which it breathed and which were used for feeding by gathering minute food particles from waters of the ocean.

Graptolites are very useful fossils to the geologist for two main reasons. Firstly, they lived floating in the upper layers of the ocean, like modern plankton, and thus became widely distributed around the world. Secondly, they evolved quite rapidly, so that most species have a relatively short duration in geological time. The combination of these two factors makes graptolites ideal for *correlating* rocks. If a species found in New Zealand rocks is known to occur in rocks of a certain age in Australia, America, Siberia, and Europe, we assume as a first guess that the age also applies to the New Zealand rocks. *Nemagraptus gracilis,* for example (shown in Fig. 5), is known in Australia, North America, Britain and Europe as well as in New Zealand, and defines a world-wide fossil zone known, for obvious reasons, as the Zone of *Nemagraptus gracilis.*

Graptolites are generally preserved in black shale and slate. In Cobb Valley they are abundant in a band of Leslie Formation strata that extends from Waingaro Peak, south across the valley floor, up over the ridge of Chaffey Stream to the ridge south of Chaffey Stream, and on to the northern end of Kakapo Spur (Fig. 19). The best outcrops for collecting fossils are on the south ridge of Waingaro Peak and the ridge on the south side of Chaffey Stream, but graptolite-bearing boulders are fairly common in Chaffey Stream itself. The graptolites are preserved as a white film and are best found by splitting slabs of black shale along their bedding planes with a chisel-bladed hammer. On many slabs the graptolites have been distorted by stretching or shortening during folding of the rock sequence.

Over 40 species are known in the rocks of Cobb Valley. Some examples are illustrated in Figure 16. Graptolites are especially important in that they provide us with almost the only means of fossil-dating the rocks of the Western Belt. In the Central Belt where trilobites and conodonts are more important for dating, graptolites are known at only one locality in the Cobb area—on the Peel Range, 2 km south of Mount Mytton. Here, however, they are generally poorly preserved and difficult to recognise.

STRUCTURE OF THE PALEOZOIC ROCKS

The geological structure—that is, the pattern of faults, folds, and deformational fabric of the rocks—of the Central Belt sequence is as complex as any part of New Zealand, and we can only 'brush the surface' here. There have been several periods of folding; the rocks have generally been tilted to moderate or steep angles and in some areas are completely overturned. Later periods have folded the folds formed during earlier periods so that

GRAPTOLITES
ANCIENT PLANKTON

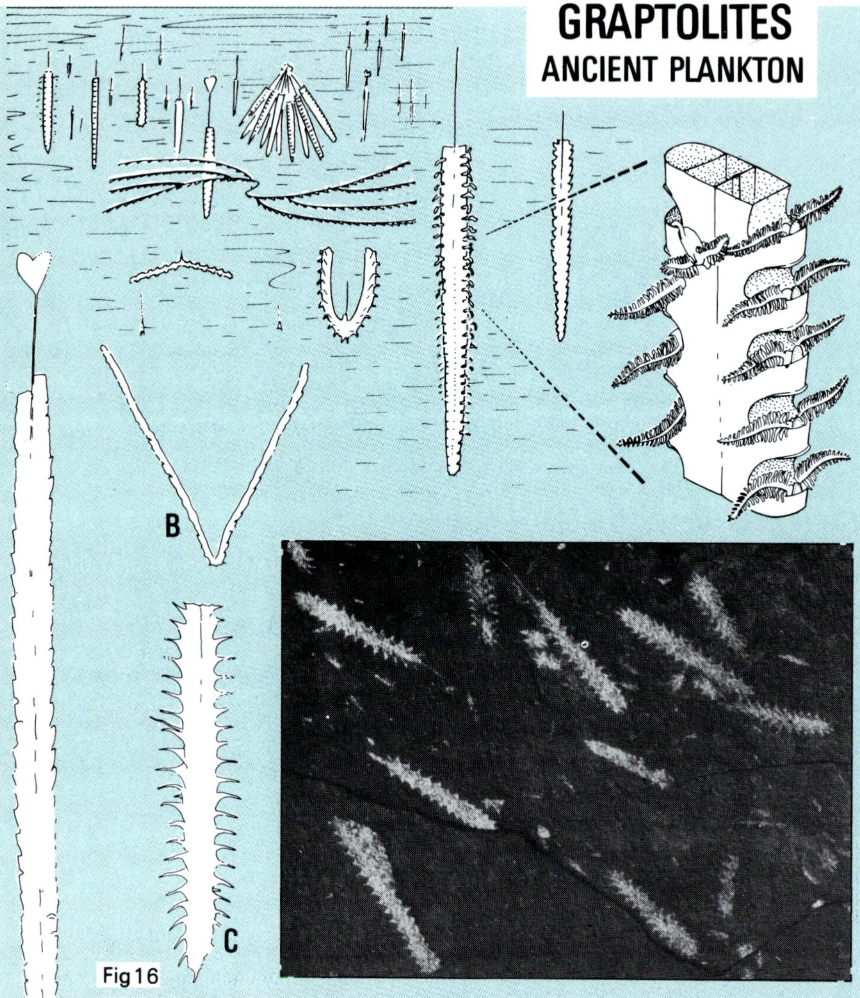

Fig 16

Most graptolite species lived suspended in the ocean water near the surface. They were colonial animals with a separate polyp (graptozooid) living in each cell (or theca) of the colony. A variety of different colony forms are depicted in their life positions. The enlarged reconstruction at right (after a drawing by P. Crowther) is based on a comparison of graptolites with their nearest living relatives, the hemichordates - an obscure group of tiny encrusting organisms. The reconstructed species is *Climacograptus typicalis.*

The drawings at left are of Cobb graptolites 2.5 times actual size. A. *Diplograptus decoratus*; B. *Dicellograptus* sp.; C. *Glossograptus hincksi.* The photograph above, D, is of a slab of black slate bearing several specimens of *Cryptograptus schaferi*, natural size.

we end up with refolded folds. When this happened again during a third period of folding the pattern becomes very complex. On top of this, we have had several periods of faulting. An early period of thrust (low-angle) faulting was followed by later periods of high-angle faults which have chopped the sedimentary sequence up into slices and moved them about—this makes it extremely difficult for the geologist to trace the original folds. The faults have moved apart rock formations that were originally together and have brought together rocks that were originally far apart. The largest and most important fault is the Anatoki Thrust (Fig. 19) which separates the Central and Western Belts and along which a great amount of movement must have taken place. A second major fault, the Devil River Fault, separates the Central and Eastern Belts (Fig. 19).

The structure is quite different in the Western Belt, where north-south-trending folds and faults can be traced over tens of kilometres and individual beds, such as the Leslie Formation quartzites, over even greater distances.

The folding and faulting movements began during the Early Paleozoic sedimentary cycle and ended in the Middle to Late Devonian, when huge masses of Karamea Granite were emplaced during an event termed the Tuhua Orogeny.

During their long history of burial, uplift, faulting and folding, the Early Paleozoic rocks have been heated and metamorphosed, changing their internal composition. Generally, metamorphic effects are not marked in the Cobb area, but in the eastern part of the Central Belt the rocks are more strongly metamorphosed and are very strongly deformed, producing the Waingaro Schist. The original rock formations from which the schist was formed can still be recognised, so the schist is shown on the map as a pattern superimposed on the other formations (Fig. 19). The Tuhua Orogeny brought to a close the early Paleozoic chapter in the Northwest Nelson geological history.

Fig. 17 Prominent rib of Summit Limestone ('The Gendarme') on the west side of Mount Benson marking the Anatoki Thrust (see above) extends north over the western shoulder of Mount Snowden, in background (arrowed). Photo: RAC.

THE ULTRAMAFIC ROCKS

One group of rocks that we have not yet discussed is the *Cobb Igneous Complex,* a distinctive belt of mafic and ultramafic rocks passing beneath the Cobb Dam, the asbestos mine, and Broken Bridge on the Old Upper Takaka Track (see map). The word 'mafic' is derived from MAgnesium and iron (FerrIC), two elements which are present in abnormally large proportions in some of the rocks. The iron produces a characteristic rusty red colour on the surface and in the soil. (Red soils are also found elsewhere—for example, on Iron Hill, and parts of the Tableland, where they develop on volcanic rocks rich in iron.) The ultramafic rocks, abnormally rich in the two elements, are mainly peridotite, pyroxenite, and serpentinite—the slippery dark green sheared rock that appears on Cobb Road half-way between the dam and the hilltop. Alteration of serpentinite has produced talc, magnesite, and asbestos at several localities but is best seen at the old asbestos mine. The origin of the ultramafic rocks is still rather uncertain, but it has been suggested that they accumulated as a layer in the Earth's crust below the Devil River Volcanics. The other possibility is that they represent a fragment of the ancient ocean floor, an origin we believe to be likely for the much larger, but unrelated, ultramafic belt of Nelson which forms Dun Mountain and the Red Hills.

The peculiar chemistry of the ultramafic rocks is reflected in the soils that develop on them, and this in turn affects the vegetation. The imbalance of elements such as nickel and cobalt makes the soil unfavourable for the natural beech-forest cover, and the ultramafic belt can be clearly defined from a distance by the sparse bush cover (Fig. 18).

Fig. 18 The effect geology can have on vegetation is well shown in this photograph taken from the magnesite quarry 1 km north of Cobb Dam. The belt of ultramafic rock indicated by the light colour pattern is covered only by scrub and, in the foreground ridge, is eroding badly. Normal beech forest covers other rock types. Photo: WStG.

WHAT WAS THERE BEFORE THE OLDEST ROCKS?

A good question, and one to which geologists still do not know the answer. We know that there must have been an older land mass to supply the sediments from which Cambrian and Ordovician strata were formed. As mentioned previously, the source area for sediments of the Western Belt is thought to be Australia and Antarctica, but the source for sediments of the Central Belt is much less certain. Neither belt seems to have been deposited directly on oceanic crust, and it is most likely that both were deposited on thin continental crust, such as that underlying the Campbell Plateau off the southeastern South Island today. Fragments of this old Precambrian continental basement are probably represented by certain metamorphic rocks called gneiss. These rocks at Charleston and in the Victoria Range near Reefton have produced the oldest radiometric dates (680 million years) and are believed to be the oldest rocks in the country. It is intribuing that some of these gneisses were formed by strong metamorphism of sedimentary rock, thus pointing to a yet older period of sedimentation and raising the question of an even older source area for these sediments. But now we are well and truly lost among the mists of Precambrian time.

THE TAKAKA ASBESTOS DEPOSIT

Asbestos has been known to occur in the upper Takaka Valley since the last century, but it was not until the prospecting activities of H.F. Chaffey in the 1930s and geological and geophysical investigations by DSIR scientists in 1939-1945 that the extent and quality of the deposit became known. Mining began in 1949 and continued on a fairly small scale until the end of 1963, when Lime and Marble Ltd took up the lease from the Hume Pipe Company. Further exploration by diamond drilling revealed two low-grade ore zones: The Pit Fibre zone contains 424 000 tonnes averaging 3.6% cross-fibre, and the Western Fibre zone, 510 000 tonnes, averaging 2.9% cross-fibre. Insufficient tonnage of quality long-fibre asbestos is present to warrant a large-scale mining operation. However, there is at least 20 million tonnes of low-grade, short-fibre asbestos which could become economic if problems caused by the health hazard of asbestos could be overcome.

The asbestos ore is found in three main types. In the first, cross-fibre veins up to 75 mm (3 inches) wide are arranged concentrically around irregular bodies of unsheared serpentinite a few metres in diameter, and surrounded by crushed serpentinite. Spectacular specimens of light yellow-green long cross-fibre were found, and it is this type of material that was worked by the early prospectors and taken out on horseback. The second type of ore is slip-fibre and is abundant in the shattered serpentinite; although it can be several centimetres long, it is brittle and of little commercial value. The third type is short cross-fibre; it is found as veins in shattered serpentinite and comprises the main ore body that was worked in the main pit.

Asbestos is the name given to a group of fibrous silicate minerals, the most important of which is chrysotile, one of several serpentine minerals. Serpentinite is a rock composed of serpentine, a hydrated magnesium silicate formed by the alteration (serpentinisation) of pre-existing rocks, particularly the ultramafic rocks dunite and peridotite, both of which are rich in magnesium silicate minerals and are present in the Cobb Igneous Complex.

Other alteration products of ultramafic rocks are talc-magnesite and quartz-magnesite, formed by dehydration and carbonation of serpentine. About 500-1000 tonnes is extracted annually by Lime and Marble Ltd from their magnesite quarry downstream from Cobb Dam. Magnesite is used as a magnesium fertiliser, as a filler, and for pottery saggers.

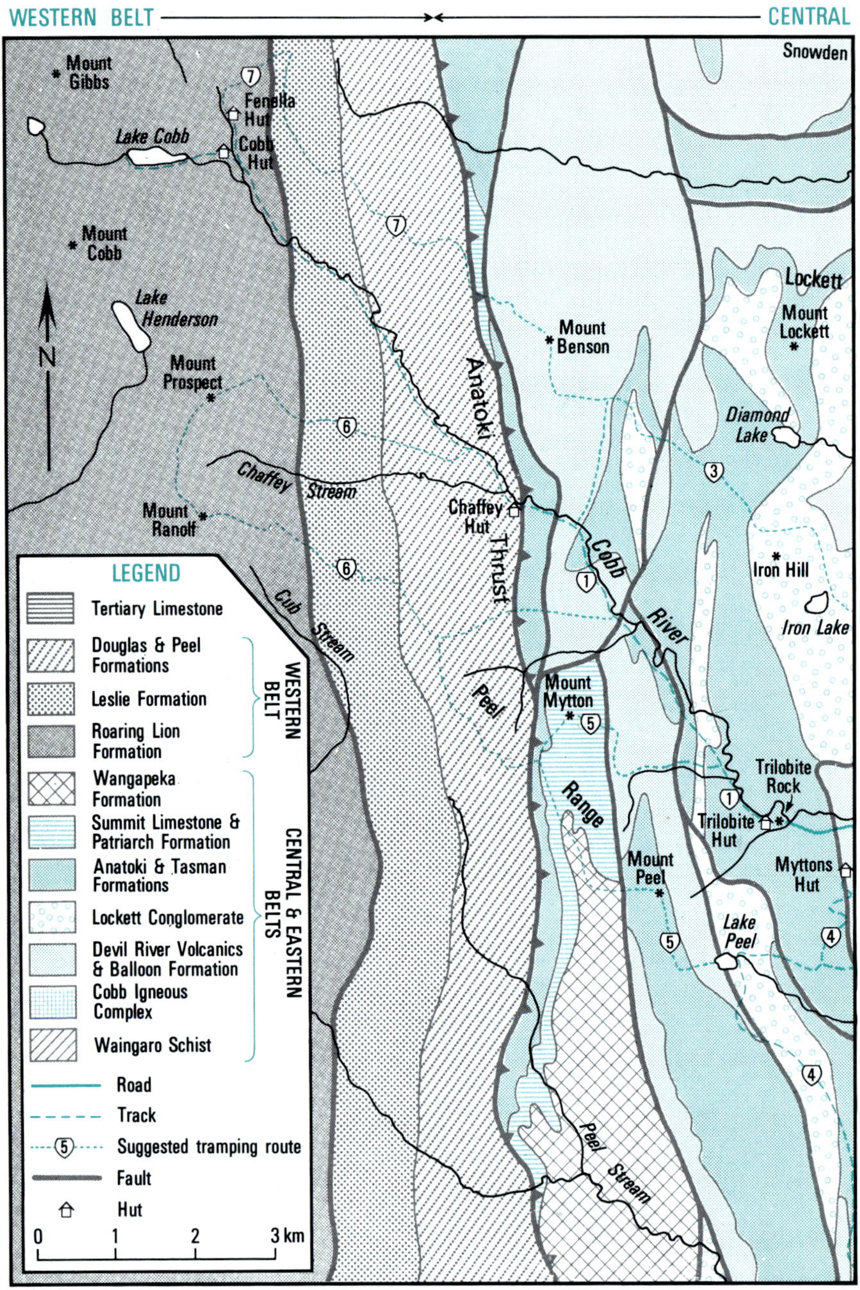
WESTERN BELT
CENTRAL
Snowden
Mount Gibbs
Fenella Hut
Lake Cobb
Cobb Hut
Mount Cobb
Lake Henderson
N
Mount Prospect
Chaffey Stream
Mount Ranolf
Cub Stream
Anatoki Thrust
Chaffey Hut
Mount Benson
Lockett
Mount Lockett
Diamond Lake
Iron Hill
Iron Lake
Cobb River
Peel Range
Mount Mytton
Trilobite Rock
Trilobite Hut
Myttons Hut
Mount Peel
Lake Peel
Peel Stream
1
3
4
5
6
7
LEGEND
Tertiary Limestone
Douglas & Peel Formations
Leslie Formation
Roaring Lion Formation
WESTERN BELT
Wangapeka Formation
Summit Limestone & Patriarch Formation
Anatoki & Tasman Formations
Lockett Conglomerate
Devil River Volcanics & Balloon Formation
Cobb Igneous Complex
Waingaro Schist
CENTRAL & EASTERN BELTS
Road
Track
Suggested tramping route
Fault
Hut
0
1
2
3 km

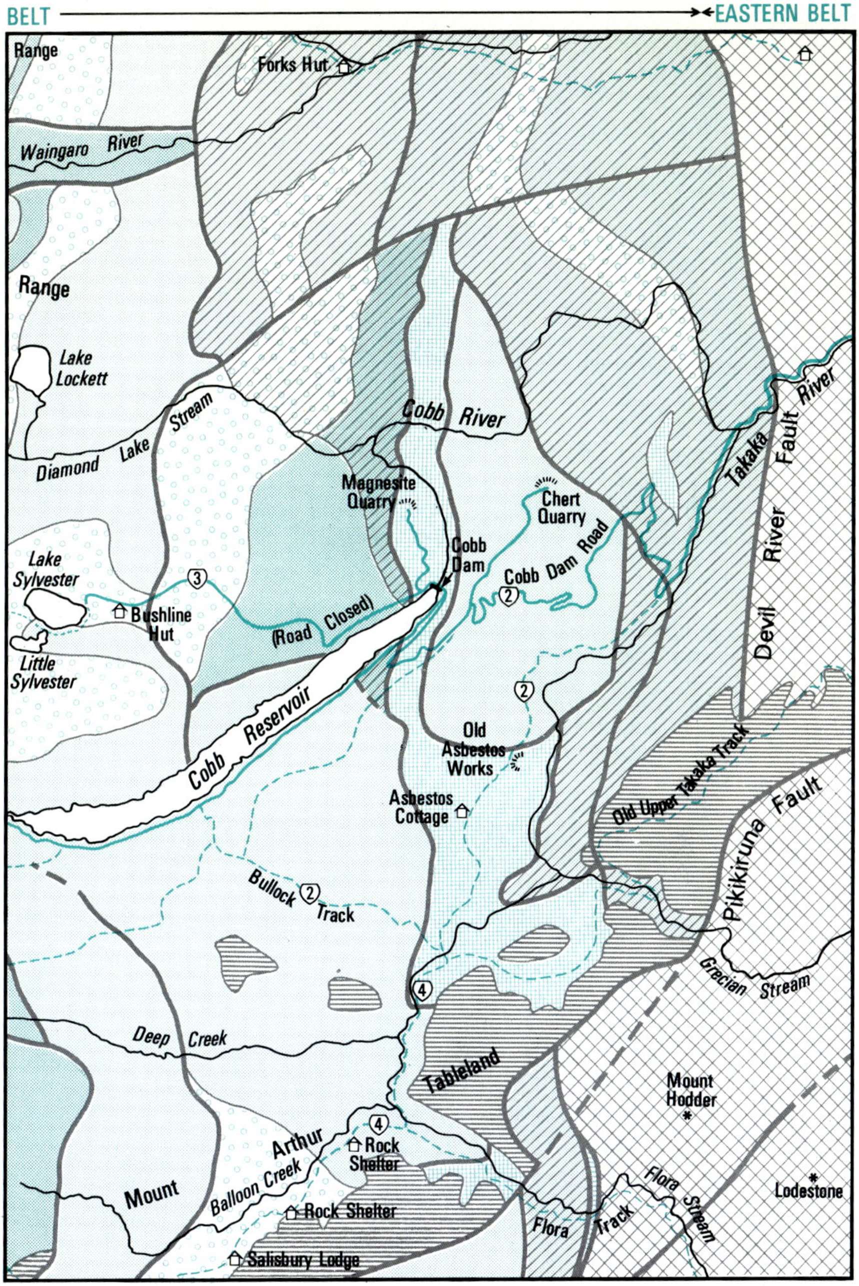

Fig. 19 Geological map of Cobb Valley area with tracks and suggested tramping routes.

Fig. 20 (top) For the first 27 km of its course, the Cobb River follows a broad, deep, glaciated valley, which at 800 m dramatically narrows. For the last 11 km the river flows down a precipitous gorge, falling 600 m. What is known as Cobb Valley is the upper, glaciated portion, part of which is now occupied by the Cobb Reservoir. This view is looking up Cobb Valley from Myttons Hut, showing the characteristic smooth-sided U-shape. Roches moutonnées (Fig. 25) project from above the valley floor to the right of the photo centre. The limestone bluffs of Mount Mytton show on the skyline at left; the prominent peak right of centre is Mount Benson. Photo: WStG.

Fig. 21 (bottom) An outcrop that tells a story, 0.5 km down the road to Cobb Dam from the ridge top. At far right is strongly weathered and leached Paleozoic rock (A), indicating a past warm climate. Overlying it is a thick mantle of solifluction debris (B), above which is about 2 m of moraine (C), itself comprising an upper layer with yellow (weathered) clay matrix and a lower layer with grey (unweathered) matrix, indicating a past cold (glacial) climate. The deposits thus represent both the warm and cold extremes of a Pleistocene climatic cycle. Loose scree material (D) covers much of the face. Photo: WStG.

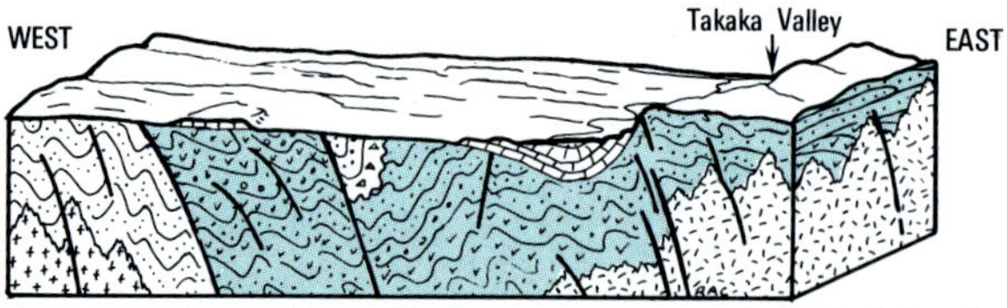

a) Present day. Resurrected peneplain; Tertiary sequence preserved in Takaka Valley, a faulted depression.

b) Tertiary. Peneplain (erosion surface) is being slowly submerged and limestone deposited.

c) Cretaceous. Permian sediments have been folded and Separation Point Granite emplaced (in east).

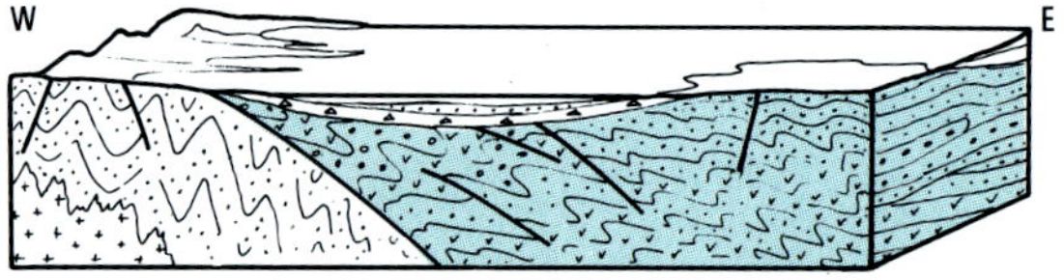

d) Permian. Erosion of early Paleozoic rocks and deposition in shallow sea.

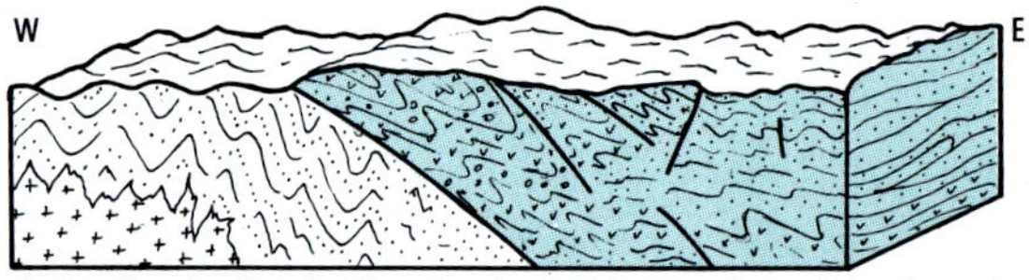

e) Late Devonian. Tectonic deformation of early Paleozoic sediments (Tuhua Orogeny) and emplacement of Karamea Granite in West.

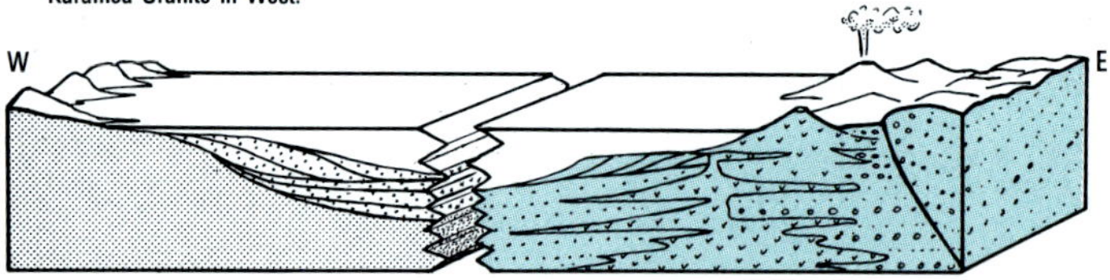

f) Cambrian–Ordovician. Western Belt sediments (left) accumulated off an active continental margin whereas Central and Eastern Belt rocks (right) formed in an off-shore volcanic region. The two regions were some distance apart.

Fig. 22 Diagrammatic cross-sections (much simplified) to show stages in the geological development of Northwest Nelson.

CHAPTER 3—THE NORTHWEST NELSON PENEPLAIN

EVENTS BEFORE THE PENEPLAIN

To understand what happened in the west Nelson region after the close of the early Paleozoic cycle, which ended in the Tuhua Orogeny, we need to look outside of the Cobb Valley region. We know there was a further submergence of the land and formation of marine and non-marine sedimentary rocks (sandstone, siltstone, and conglomerate) in the Permian Period (late Paleozoic). This Permian sequence was subsequently folded and metamorphosed in a second orogeny, or mountain-building cycle, during which more granites were emplaced; these younger granites are mainly those of the Separation Point Batholith but include isolated smaller bodies such as the Onahau Granite near Takaka. The only remaining vestige of the once extensive Permian sedimentary sequence is preserved as an infolded strip at Parapara Peak in the northern Tasman Mountains. The second orogeny, the Rangitata Orogeny, has 'overprinted' the effects of the earlier Tuhua Orogeny in the Golden Bay region where high-temperature metamorphism has produced a wide belt of biotite schist. In the Cobb Valley area, there appears to be little metamorphic effect from this orogeny; there is likely to have been some folding and faulting of the lower Paleozoic rocks but again, the effects as preserved in the rocks do not appear to have been very marked.

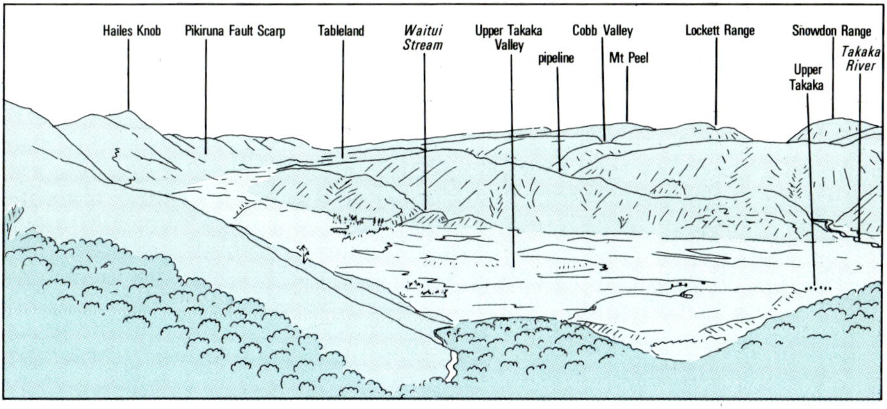

Fig. 23 Takaka Hill road (State Highway 60) affords an excellent view of the Northwest Nelson peneplain. This sketch, from about 1.5 km down from the top of the road, shows the Pikikiruna Fault Scarp bordering the Pikikiruna Range (formed of Paleozoic rocks) at left rising above the uplifted peneplain surface, which is defined by the gently dipping flat topped ridges. The Upper Takaka Valley is a tectonic depression formed by downwarping of the Tertiary sequence (and peneplain surface). The road, descending from the top of Takaka Hill, passes through Paleozoic limestone and marble to the first hairpin bend, then through quartzite and schist (Paleozoic) to the 2nd hairpin, through more marble and into Tertiary limestone (with fossils) at the 3rd hairpin, or Eureka Bend.

THE PENEPLAIN

Well after the close of the Rangitata Orogeny, in late Cretaceous and earliest Teritary time (70 million years ago), coal measures, including coals of the Puponga Coalfield, were deposited in the Wakamarama Range area. These sediments include boulder-bearing conglomerates derived from the Paleozoic basement. They lie in steep-sided valleys and depressions cut into the Paleozoic basement and undoubtedly formed in a land of moderate relief. During a later period, but still in the early Tertiary, more coal measures were deposited widely across northern and western Northwest Nelson and rest on a flat erosion surface generally underlain by lower Paleozoic rocks; in Wakamarama Range the erosion surface has been cut over the older coal measures. Thus between the times of deposition of the older and younger coal measures (i.e., about 60 million years ago) the hills were reduced by erosion to a lowland plain termed a *peneplain* (*pene* = almost; *plain* = level surface).

During the erosion of the peneplain, the rocks were deeply weathered. Evidence from fossil plant pollen and marine molluscs indicates that the climate in New Zealand during the early Tertiary was considerably warmer than that of today—equivalent to that of the subtropics. As in the subtropics today, the soils were strongly leached, forming an iron pan and remnants of thick iron pan deposits that were formed during the early Tertiary are found near Onekaka where they were once mined for iron ore (limonite). Clay deposits formed during the same weathering period have been largely removed by erosion, but remnants can be found at such places at Puramahoi where they are quarried for the ceramics industry and craft pottery.

THE TERTIARY STRATA

Submergence of the land followed deposition of the coal measures, and once more marine sediments were deposited. Limestone and calcareous sandstone of mid Tertiary age were deposited widely over Northwest Nelson and, indeed, over much of New Zealand. The beds are worked for cement manufacture at Tarakohe; they also crop out in the lower and upper Takaka Valley and at Eureka Bend on Takaka Hill Road. They are richly fossiliferous and contain brachiopods, bryozoans, and molluscs, including the large fan shell *Athlopecten athleta.* On the Tableland, gently-dipping remnants of the once-extensive limestone sheet can be seen from the headwaters of Leslie River northeastwards to Grecian Stream and further on to Upper Takaka. Huge fallen blocks of Tertiary limestone clutter the Takaka River from the Flora Stream junction down to the Bullock Creek Track junction at Broken Bridge, and the rock shelter near Salisbury Lodge is a natural overhang in Tertiary calcareous sandstone and limestone.

Overlying the limestone are the youngest marine beds—mudstone (Tarakohe Mudstone) and sandstone—and because they are soft and uncemented, they are readily eroded. The nearest remnants are found at Upper Takaka.

Fig. 24 Aerial view of the Mount Arthur Tableland. Balloon Creek, in the left foreground, meanders across the exhumed peneplain surface. Bush covers Tertiary limestone resting on the peneplain surface in middle ground, beyond which is the Karamea Fault and Mount Arthur Range (with conspicuous outcrops of limestone and marble). The Moutere Depression and Nelson ranges can be seen in the distance. Photo: DLH.

RESURRECTION OF THE PENEPLAIN

Just when the sea retreated and formation of the present landscape began is uncertain, but we know it must have been after deposition of the youngest marine beds in the late Tertiary, about 10-15 million years ago. As the land was uplifted, it was gently arched, and major fault scarps began to form along the Wakamarama and Pikikiruna faults, bounding the Takaka and Aorere valleys. For their lower 25 km, the Takaka and Aorere rivers thus flow down depressions which were partly formed by faults and which, therefore, are not entirely the work of river erosion. The full Tertiary sequence is preserved in these depressions—but elsewhere only remnants remain. The old peneplain surface was once more exposed to erosion, and streams that had cut easily into the soft Tertiary cover sequence suddenly found themselves flowing through hard Paleozoic basement rocks. The rather meandering courses of the river valleys that formed on the gently-arched Tertiary sequence were 'inherited' by the rivers that cut into the basement rocks and eventually became the rivers of today. Because the Paleozoic rocks are much more resistant to erosion than the softer Tertiary cover sequence they have been slower to erode. For this reason, the relatively flat surface of the resurrected peneplain can still be clearly seen and is represented by the flat surface of the tableland, and flat-topped ridges such as that leading from Mount Peel down to Balloon Hut. Farther west, in the central Tasman Mountains, all traces of the flat surface have been eroded away, but its former existence is betrayed by an accordance of summit heights of the highest peaks. All these features can be clearly seen on a fine day from the Takaka Hill Road (State Highway 60) at the top hairpin bend, or from the top of Mount Peel.

CHAPTER 4—GLACIATION

The Ice Age, or Pleistocene Period, is now believed to have begun about 1.6 million years ago. There are several theories as to what triggered the Ice Age; one suggests a decrease in the amount of solar radiation reaching the Earth's surface because of an increase in upper atmospheric dust produced from a world-wide increase in volcanic activity. Although the theories are still being debated, it is probable that no one factor alone caused the onset of glaciation but rather a combination of several, probably unrelated, events. Ice ages are rare in the history of the Earth.

The Ice Age has seen a succession of climatic oscillations—cold (glacial) periods when glaciers form in mountain valleys and the polar ice caps grow, followed by warm (interglacial) periods when glaciers retreat up their valleys (as in the Southern Alps) or disappear altogether, giving away to the returning forest (as in Northwest Nelson), and polar ice caps shrink in size. For many years it was believed that there were four or five main oscillations, based on the study of glacial moraine deposits dumped by the spreading glaciers across Europe and North America. But now, thanks largely to isotope studies of sea-floor sediments, it is known that there have been at least 23 oscillations in the last one million years alone. The duration of cold periods ranges from 16 000 to 60 000 years, and the intervening warm periods have a similar duration. At present we are in a warm period that has been going for about 11 000 years. Unfortunately, it is not yet possible to predict when the pendulum will swing back and the next cold period will begin, but this is not expected to happen for at least several thousand years. The oscillations are caused mainly by fluctations in the amount of the Sun's radiation received by the Earth, and their magnitude and duration are affected by other factors such as the wobble in the Earth's axis of rotation.

One can imagine the drastic effect these oscillations must have had on the vegetation. In the natural vegetation of the west Nelson region, mixed podocarp-broadleaf forest (with rimu, totara, rata, and beech) occupies a lower altitude zone than that of beech forest (composed almost entirely of beech). All the vegetation zones have migrated rapidly to and fro with the climatic changes, but because beech forest cannot migrate as rapidly as podocarp-broadleaf forest, there are places such as Baton Valley where the broadleaves have leapfrogged over the beech and now occupy a higher level. During the cold periods some warm-climate plants became extinct in New Zealand although the warm climatic conditions they require have now returned; for example, the large-leaved beech species (the *Nothofagus brassii* group) which today grow in New Caledonia and New Guinea were formerly important members of the New Zealand forest.

Another effect of the oscillating climate and fluctuating size of the polar ice caps was a corresponding rise and fall of sea level. During the peak of a cold period much of the world's surface water was locked in the polar ice caps so that sea level fell to about 100-150 metres below its present level. Much of the continental shelf beneath the coastal waters became land, greatly increasing the New Zealand land area. The North and South Islands were joined by a broad land bridge between Taranaki and Northwest Nelson, enabling the migration of plants and animals.

GLACIAL FEATURES IN THE COBB VALLEY

The most obvious effect of glaciation is the form of the valley itself. The relatively straight course of the valley above the dam, its smooth sides and broad open U-shaped profile are characteristic of glacial valleys and result from ice abrasion and scouring. In contrast, mountain valleys which have not been glaciated, such as the upper part of the Takaka River valley, or the Leslie Valley, have V-shaped profiles and winding courses. The maximum extent reached by the glacier is indicated by the valley form and distribution of glacial deposits, especially of moraine. (Moraine is the deposit formed when a glacier dumps its load of sediment at its terminal or along its margins and is an unsorted mixture of silt and boulders.) Good outcrops of moraine are seen as far down valley as the road from the dam to the Magnesite Quarry (Fig. 33) and indicate the limit reached by the Cobb Glacier during the Ice Age. Another cold-climate deposit is solifluction debris—small angular fragments of rock produced by freeze and thaw action, forming a drape over the valley side. One very interesting outcrop in a cutting on the road down to the dam (Fig. 21) shows a thick, layered mantle of solifluction debris resting on highly-weathered Paleozoic rock with about three metres of moraine resting on top. The moraine itself is composed of an upper layer with yellow (weathered) clay matrix and a lower layer with grey (unweathered) matrix.

The Cobb Glacier occupied the valley of an older river, the 'proto-Cobb River', as did glaciers in the valleys of the Waingaro, Stanley, Roaring Lion, Anatoki, and Burgoo rivers of the Tasman Mountains. Small tributary glaciers have left hanging valleys, a good example being that of Hannah Creek, just downstream from Chaffey Stream. High level basins, or cirques, in which the snow that fed the glaciers accumulated, are common in the Cobb area. They often have a sill or lip, behind which accumulates a small lake or tarn. Almost all the natural lakes and tarns in the area occupy cirques; examples are Lakes Lockett, Sylvester, Little Sylvester, Iron, Lillie, Ruby, Round, Camp, and Peel. Lake Cobb, Lake Henderson, and Island Lake occupy portions of a glaciated valley that were scooped out or overdeepened by the glacier.

Fig. 25 Glacial scouring and grooving by the Cobb Glacier during the Ice Age has left its mark on this protruding bedrock outcrop (Tasman Formation) in the valley floor (locality 2 in Fig. 30) near the lake head. As the glacier flowed down the valley, over the outcrop, it carved the longitudinal hollow extending along the side of the outcrop just above ground level and planed smooth the surface of the outcrop. Photo: WStG.

One interesting feature, discovered by seismic surveys for the Cobb damsite, is that the bedrock floor of the valley, beneath the gravels and moraine, deepens upstream for some distance; this part of the main valley was also scooped by the glacier and may have been occupied by a lake some time during the Pleistocene Period, before being filled by gravels and moraine. Thus the present Cobb Reservoir may occupy the site of an old Pleistocene lake. Some early geological surveys in the valley suggested that flats, such as that below the junction of Mytton Creek, may represent filled-in lakes, but as no lake sediments have been reported, we cannot verify this suggestion.

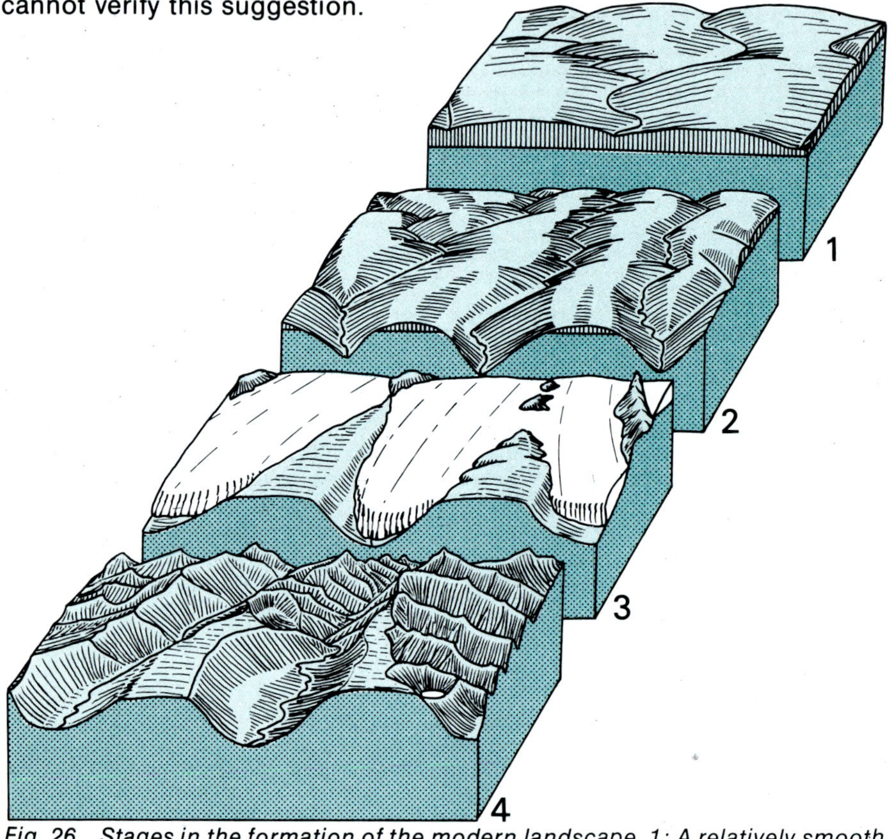

Fig. 26 Stages in the formation of the modern landscape. 1: A relatively smooth, flat land surface was uplifted in late Tertiary time and carried meandering rivers. 2: As uplift continued, the rivers cut deeper and the soft cover rocks were eroded away. Eventually the more-resistant Paleozoic basement was reached. The peneplain cut on the basement rocks was once more exposed to surface erosion. 3: With the onset of the Ice Age, glaciers filled the upper reaches of the river valleys. Glacial erosion then straightened and deepened the valleys, and cut deeply into the basement rocks. 4: After retreat of the ice about 15 000 years ago, the rivers reoccupied their valleys. The only remanants of the old peneplain surface are the flat-topped spurs and the Tableland.

As the ice passed over the valley floor, all soil and weathered surface rock was removed, leaving the bedrock scraped clean. Rock projections in the valley floor were smoothed over and became whale-backed outcrops called roches moutonnées, and some show rounded grooves carved by the glacier and its sediment load (Fig. 25). The rounded hummocks of bedrock seen in the valley floor between the head of the reservoir and the camping ground are roches moutonnées, and many other examples are found farther up the valley. Many are steepened on the down-valley side, a feature caused by glacial plucking as the ice moved down the valley.

COBB HYDRO-ELECTRIC POWER SCHEME

Development of the valley for hydro-electric power has had a long and somewhat chequered history. The initial project was undertaken by a private company in 1937, but the whole scheme was taken over by the government in 1940 and construction of an access road up the Takaka River gorge began. The original scheme employed a weir for water storage, built about half-way between the present dam and the head of the reservoir. The storage reservoir, called Lake Halley, was soon found to be insufficient, and several of the tarns feeding tributary rivers, such as Lake Sylvester, were dammed and controlled. Investigations for the foundations for a new and larger dam revealed some interesting geological information. What looked like outcrops in the valley floor turned out to be huge boulders suspended in moraine (the terminal moraine of the Cobb Glacier), and the site was thus proved unsuitable for a concrete dam. A compacted earth-fill dam was therefore built, with a concrete spillway, and the earlier weir was flooded by the new reservoir.

The present dam is 40 m high with a crest length of 220 m. A 2.6 km tunnel connects the outlet to the twin penstocks which descend the side of the Takaka River valley. When full, the reservoir is 807 m above sea level and holds 25.6 million cubic metres of water. The gross head (or height difference between reservoir and power house) is 594 m, the greatest in New Zealand, and permits the generation of a considerable amount of power from a relatively small water flow. Capacity of the generating station is 32 megawatts and the power is fed into the South Island Grid.

Fig. 27 Cobb Valley in January, 1948. The construction camp for the power scheme can be seen, as well as the newly-completed road over the hill from the generating station, and along the side of the future reservoir.

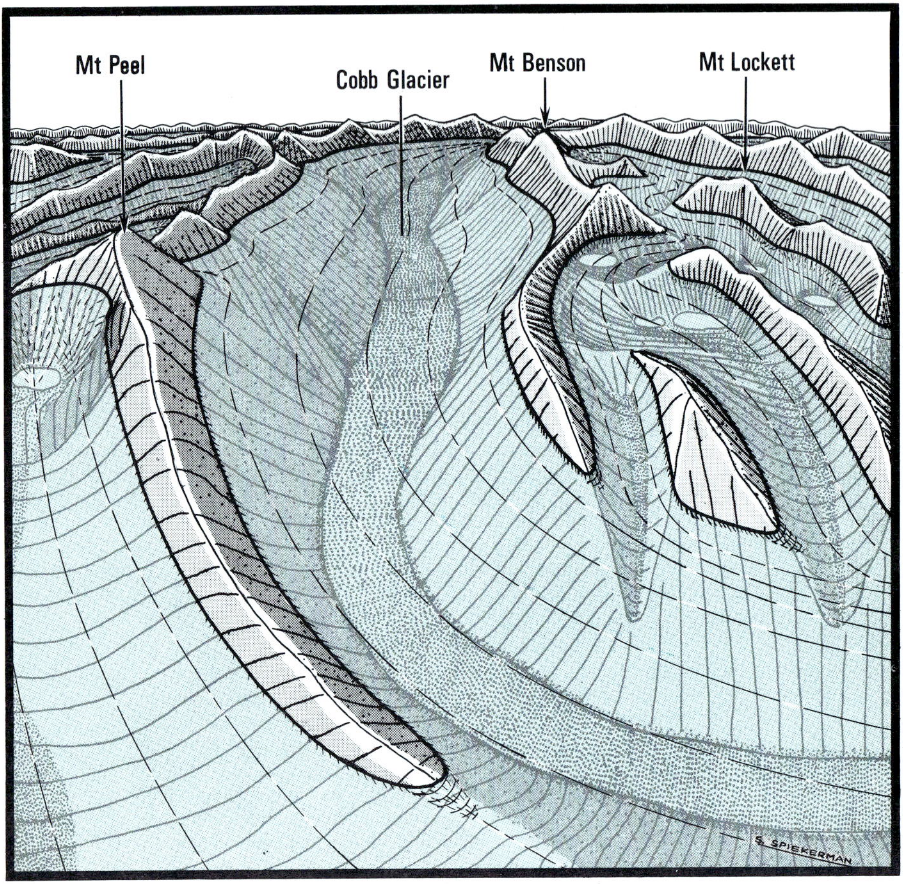

Fig. 28 The Ice Age was a period of see-sawing climatic changes. During the warm intervals the rivers flowed and vegetation covered the valleys. During the cold intervals the valleys were filled by glaciers, which in their upper reaches formed a vast ice plateau. This was pierced by peaks which now form the highest points on the divides. The figure depicts a scene at the height of an ice advance (cold period); the valley form beneath the glaciers is also shown.

CHAPTER 5—TRAMPERS' GUIDE

The Cobb—Mount Arthur—Leslie Valley region is part of the Northwest Nelson Forest Park and has many excellent bush huts and well-formed, clearly-marked tracks, all provided and maintained by the New Zealand Forest Service. Together with its impressive mountain scenery, remarkable botany and geology, the region ranks as one of the most interesting and rewarding in the country for the tramper and naturalist. Armed with the geological map (Cobb Sheet by G.W. Grindley—see "Further reading") and the one mile to the inch topographic map one can undertake numerous walks and tramps to see features of geological or botanical interest. I have suggested a few routes below and they are indicated in Fig. 19. Some are shown in the Forest Service brochure on Cobb Valley.

A WORD OF WARNING

The main hazard in roaming the mountains is being caught unprepared by a rapid deterioration in the weather. A trip that is easy in fine weather can become difficult or even dangerous if the weather closes in. It can snow at any time of the year in the mountains. Wear solid shoes and always carry windproof clothing and spare food; watch for weather changes. If you have had little experience of tramping in the mountains stick to the less-demanding routes (nos. 1, 2 and, if you are reasonably fit, 4).

Please note that the map with routes (Fig. 19) *is not* an adequate route map for those routes which do not follow marked tracks (Routes 3, 5, 6 and 7). Be sure to obtain a copy of the Lands and Survey Department one mile to the inch topographic map (S13, Cobb) which shows the rivers, tracks, roads, huts and ridges as well as contoured topography. The times indicated for each route are for return trips.

Fig. 29 The Tableland, with Mount Arthur, left; the Twins, centre on skyline; Balloon Creek on left. Photo: LL.

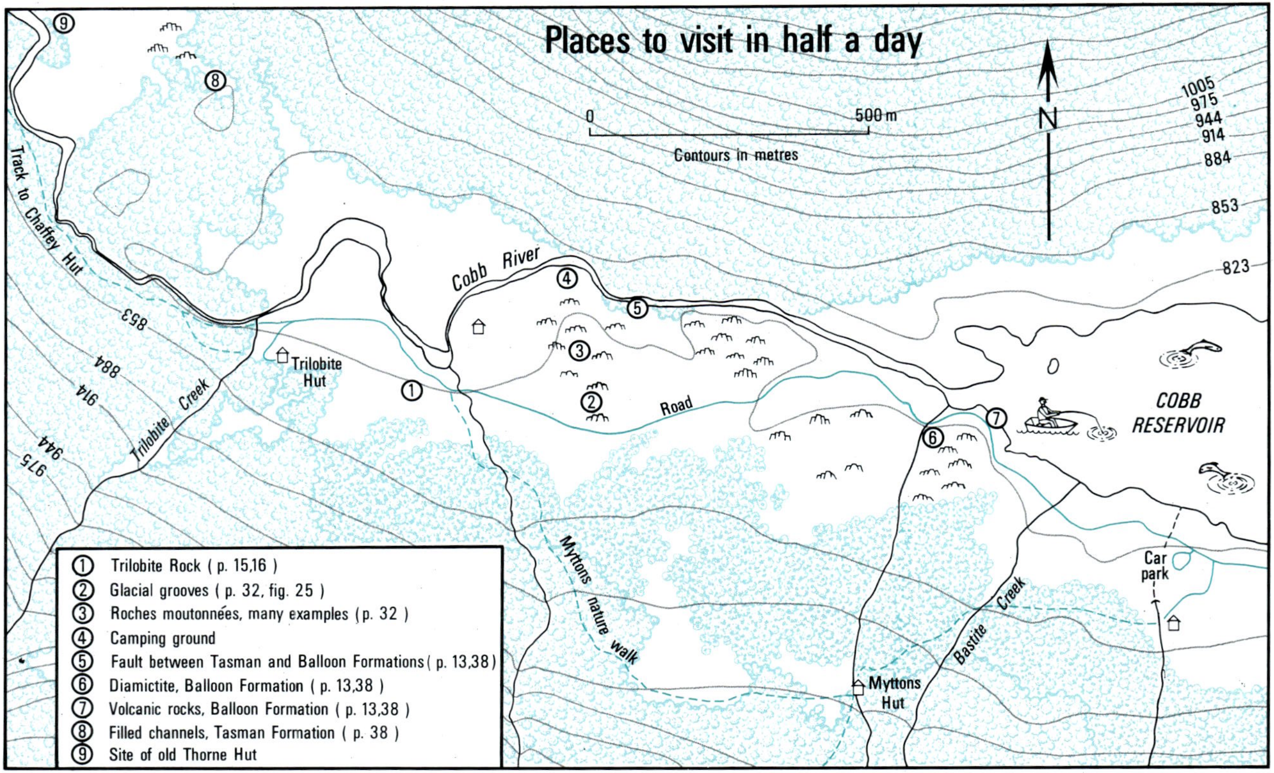

Fig. 30 Valley floor between Cobb Reservoir and Trilobite Hut, showing the location of several features of interest that you can see in half a day. In addition, you can visit the Magnesite Quarry and the chert quarry (both shown on map, Fig. 19); the chert quarry is accessible by 4-wheel-drive vehicle. Other features of interest that can be seen from Cobb Road are outlined on p. 39.

ROUTE 1—VALLEY FLOOR UP TO CHAFFEY STREAM (½—1 day)

For those who are not climbers this is your trip!

From the car park at Trilobite Hut an excellent, well-marked track leads up the valley floor to Lake Cobb and Fenella Hut at the head of the Cobb Valley. Chaffey Stream is one to two hours walk and a good target for an easy and interesting half- to one-day trip.

Starting at the head of the reservoir, the first rocks encountered are volcanic conglomerate, well-exposed in a rock ledge below the road (Fig. 30, no. 7). Next are the dark muddy diamictites of the Balloon Formation (Fig. 30, no. 6). As we approach the camping ground we pass into hard laminated siltstone of the Tasman Formation which forms low knobs (roches moutonnées) smoothed by glacier action (Fig. 25; Fig. 30, nos. 2, 3). Bedding is almost vertical. Trilobite Rock (signposted) right on the roadside is a lens of Cambrian trilobite limestone within the siltstones. It is the most richly fossiliferous of a "string" of limestone lenses in the siltstone. In Cobb River (Fig. 30, no. 5) there is a fault between Tasman and Balloon Formations. Highly contorted rocks can be seen in the southern bank when the river is low.

About 250 m past Thorns Creek another Cambrian limestone lens is encountered with rare fossils. A short diversion from here across the river (you will get wet feet) will bring you to more fine examples of roches moutonnées, unfortunately (from the geologist's point of view) partly bush-clad. The Tasman siltstones here show channels that were cut into them at the time of their deposition (Fig. 30, no. 8). The channels are filled with conglomerate and sandstone. Bedding is gently dipping. At Myttons Creek the track passes close to bluffs formed of yet more Cambrian limestone; from here the high rugged slopes of Mount Mytton can be clearly seen to the south, and a large prominent scree lies below cliffs formed of a different limestone—the Summit Limestone of Ordovician age. After Myttons Creek the track crosses an alluvial flat along which there are few outcrops of Devil River Volcanics. By crossing the Cobb River (wet feet again) one can reach a large area with numerous knobby outcrops—all of Cambrian volcanic rocks. Beyond the flats the track passes close to rapids in Cobb River with large boulders and outcrops of Devil River Volcanics.

All rocks encountered to this point are of Cambrian age, but at Chaffey Hut we pass over the largest and most important Paleozoic fault in Northwest Nelson—the Anatoki Thrust—and leave the Central Belt. Unfortunately the fault cannot be seen in the valley floor, but its position is defined by mapping the rocks on either side.

Rocks of the Western Belt appear to the west of the Anatoki Thrust and the track traverses the various formations on its way up the valley. A selection of these rocks is represented, however, by the boulders in Chaffey Stream. Black slate boulders, some of which contain graptolites, and grey quartz sandstone boulders are common.

The quaint little beech slab hut near Chaffey Stream, Chaffey Hut, was not occupied by the legendary Chaffey of Asbestos Cottage fame—it was built well after his death by Jack McBurney, a forest ranger, in 1957 (Fig. 32).

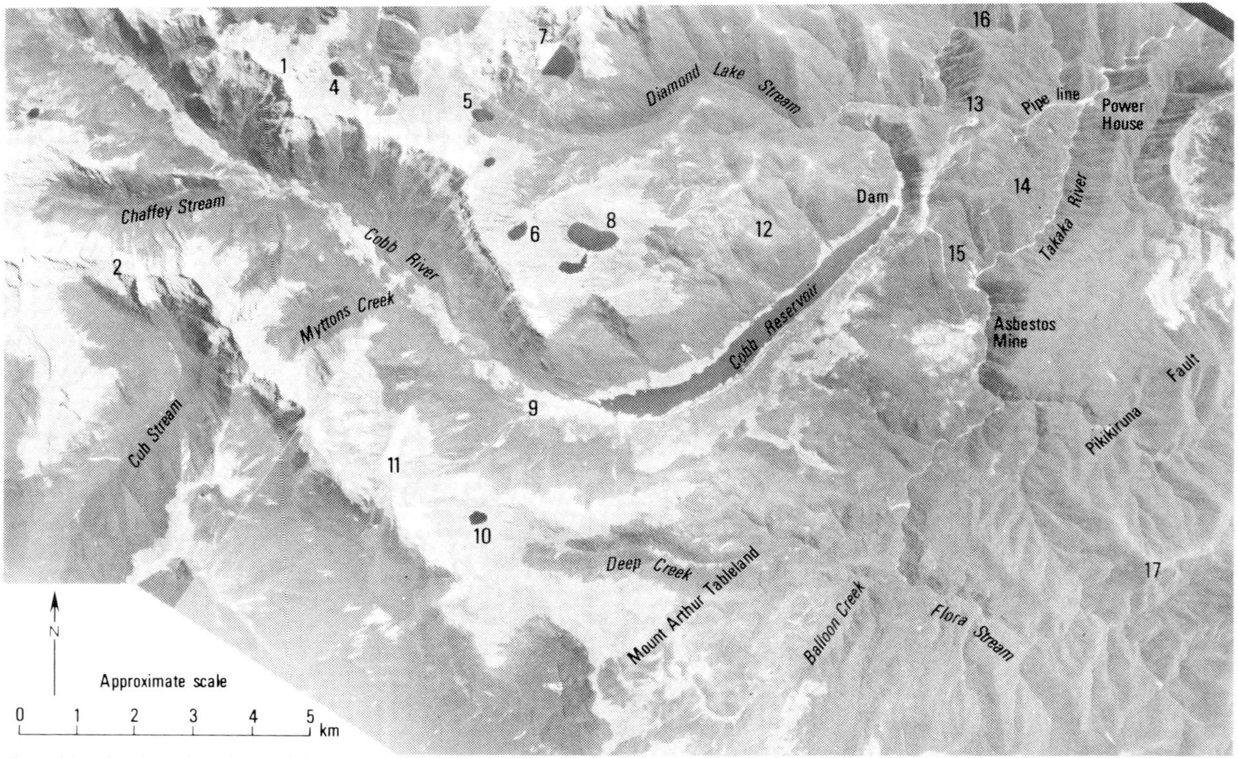

Fig. 31 Vertical aerial view of Cobb Valley and part of Tableland, taken from an altitude of 6000 m. Bush and lakes appear dark, open country light. Compiled from two photographs (Lands and Survey Department 2033, 4029/33 and 4030/45; published by permission). Key to numbered features: 1, Mount Benson; 2, quartzite bands in Leslie Formation; 3, Mount Ranolf; 4, Ruby Lake; 5, Lake Diamond; 6, Iron Lake; 7, Lake Lockett; 8, Lake Sylvester; 9, Trilobite Hut; 10, Lake Peel; 11, Mount Peel; 12, road to Lake Sylvester; 13, chert quarry; 14, Cobb Road; 15, telegraph line; 16, gorge of the lower part of Cobb River; 17, Mount Hodder.

Fig. 32 Chaffey Hut at Chaffey Creek, built of beech slabs by Jack McBurney in 1957. Photo: WStG.

ROUTE 2—BULLOCK TRACK—ASBESTOS WORKS—COBB ROAD (½-1 day)

This is a moderately easy walk and follows a formed track or road all the way.

The Bullock Track via Peat Flat (signposted) leaves the road about halfway along the Cobb Reservoir. After passing over one of the very rare, and unfortunately unfossiliferous, limestone lenses in the Balloon Formation it climbs up the south wall of the valley to a broad flat depression in the ridge crest. This flat is probably not part of the old peneplain surface—unlike the flat ridge from Mount Peel down to Balloon Hut—but probably marks the course of an old, high-level glacier, possibly a spill-over distributary of the Cobb Glacier. Fine views of the reservoir and lower valley are one reward for the steep climb. The Bullock Track, incidentally, is so named because it is the route used by drovers early this century to bring their cattle from Graham Valley and the Tableland to graze in Cobb Valley each summer.

The track then descends gently for 300 m (1000 ft) to the swingbridge at Broken Bridge on the Takaka River. There are almost no rock outcrops on the route, but the bush is lovely. In Takaka River, just upstream from the swingbridge are outcrops of banded gabbro, a basic igneous rock and part of the Cobb Igneous Complex. The track forks at the river; we follow the left-hand branch, which climbs along the north side of the Takaka River valley to Asbestos Cottage, the home of Mr H.F. Chaffey and his wife Annie for over 40 years. A fascinating account of this remarkable couple is given by Jim Henderson in his book "The Exiles of Asbestos Cottage".

Descending from Asbestos Cottage to the old asbestos works we pass through outcrops of black to blue-black ultramafic rock. They are mainly pyroxenite, peridotite, and serpentinite. Sheared serpentinite is a conspicuous green colour and readily breaks into fragments, causing slips. The asbestos is found as veins, generally only a few millimetres thick, in dark serpentinite, and specimens can still be collected from the old pit. The road from the asbestos works out to the Cobb Road is now traversable only by foot; outcrops along it are mainly of gabbro and Devil River Volcanics.

The Cobb Road climbs through more volcanics and sheared rocks of the Balloon Formation to the top of the ridge above the dam. Short detours enable one to visit the tunnel portal and twin penstock for the Cobb Hydro-electric Power Station. From the ridge-top fine views of the Mount Arthur Tableland and Mount Arthur Range can be seen. The peneplain surface is well displayed and, in fact, at this point you are standing on it. Outcrops of bright green sheared serpentinite are exposed in road cuttings on the way down to the dam and, a little further on, an interesting outcrop which reveals details of the glacial history of the valley is seen (see Fig. 21).

ROUTE 3—LAKE SYLVESTER—MOUNT BENSON (1 day)

This trip involves climbing 700 m (2300 ft) and is suitable for the experienced tramper. It is a fine-weather trip only, and you will have to keep moving. Most of the route is untracked.

Starting from the dam, a 4-wheel drive road (not open to public vehicles) runs along the north side of the reservoir then climbs up through the bushed valley side and comes out onto the open tops, ending at Lake Sylvester. At a culvert, 1 km from the dam, is a good outcrop of orange-grey laminated siltstone of the Anatoki Formation. Small faults and evidence of disturbance of the sediments before they were consolidated can be seen in the outcrop. In the outlet channel of Lake Sylvester, dark sulphide-bearing rocks of the Balloon Formation can be seen. Climbing up to Iron Lake and then up to the Iron Hill ridge beyond, we pass among numerous rock faces of one of the most distinctive and easily recognised rock formations in the region—the Lockett Conglomerate. Two sets of fracture cleavage at right angles cause the conglomerate to break into huge blocks. Outlines of rounded boulders and pebbles of which the conglomerate is made are clearly seen in the fractured faces. From the ridge crest one has a magnificent view taking in Nelson, Tasman Bay and the Richmond Range, the Spenser Mountains, Mount Owen, Mount Kendall, and the western and northern Tasman Mountains. The Mount Arthur Range and Tableland are clearly seen and the remnants of the peneplain surface are obvious. To the north and northeast the peneplain has been completely eroded away but the accordance of summit heights and of ridges descending to the Takaka Valley can be seen. On exceptionally clear days the Kaikoura Ranges can be seen. The basins containing Lakes Sylvester, Little Sylvester, Iron Lake and Lake Lillie are good examples of glacial cirques, and Lake Lockett to the north is a perched cirque or hanging valley.

The walk along the ridge crest from Iron Hill to Mount Benson passes from Lockett Conglomerate into Tasman Formation siltstone, with three conglomerate bands followed by a Cambrian limestone lens. It then passes through Devil River Volcanics, in which coarse black tabular pyroxene crystals can be seen, into more Tasman Formation slate. Next, a prominent band of Cambrian limestone in the Tasman Formation crosses the ridge and descends into Cobb Valley and is a good point to end the traverse. The quickest way home is down into Cobb Valley and back along the track to the reservoir road.

Fig. 33 Large boulder weathering out of moraine on cutting in road to Magnesite Quarry. Photo: WStG.

ROUTE 4—LAKE PEEL—TABLELAND—BULLOCK TRACK (a long day)

A long but not difficult walk with the big climb (400 m) at the beginning; this trip should not be attempted in conditions of low cloud or mist except by experienced trampers or those who know the area well. The route follows a marked track all the way.

From the Myttons car park the track follows the Myttons bush walk to Myttons Hut. A fine view up and down valley from here shows the smooth-sided glacial valley profile well. The track to the ridge top is well graded, and from the ridge top there are further impressive views of the valley topography, Mount Arthur and the Tableland. The track then sidles into Lake Peel, which occupies a cirque passing through Lockett Conglomerate. More conglomerate, then volcanics, and the track comes out on the crest of the flat-topped ridge leading from Mount Peel down to the Tableland, a remnant of the old peneplain surface. Looking across towards the Mount Arthur Range, remnants of the Tertiary limestone cover on the peneplain surface are bush-capped. A short detour (which you haven't really got time for) to the southwest slope of the ridge reveals two long-disused water races, built to carry water from a small dam to the Balloon Diggings. The walk to Balloon Hut and Salisbury Lodge follows the gently-dipping peneplain surface and passes through Tertiary limestone at Cundy's Creek, where the tunnel-like Bishop's Cave is developed, and (with a short detour) the rock shelter, a natural overhang, beyond Salisbury Lodge. Gold worked at the Balloon Diggings, Cundy's Creek and elsewhere on the peneplain surface is all recycled; it came originally from quartz reefs in the Paleozoic rocks but was eroded and deposited by the early Tertiary rivers that cut the peneplain surface itself and was subsequently preserved in gravels and gritty limestone at the base of the Tertiary strata. The gold-bearing gravels and grits have been re-exposed by the present weathering cycle and the gold is thus now being recycled. From Salisbury Lodge the track sidles down Balloon Creek, past the imaginatively developed Growler Rock Shelter, to the swing bridge across Flora Stream where it meets the Flora Track. At the stream junction the river becomes the Takaka and the track sidles along the eastern bank through dark sandstone and mudstone of the Balloon Formation. Huge blocks of Tertiary limestone detached from the cliffs above litter the hillside and river bed and, through the bush, one can see high up on the opposite bank the white cliffs formed by more Tertiary limestone. At Broken Bridge (now a perfectly good suspension foot bridge) the track is joined by the Old Upper Takaka Track—for many years the main access route to the asbestos deposits. At this point we are standing on rocks of the Cobb Igneous Complex, and the return trip home via the Bullock Track is covered in Route 2.

ROUTE 5—LAKE PEEL—PEEL RANGE—MOUNT MYTTON (1 day)

Quite a bit of climbing in the open tops and no marked track for most of the way—suitable for experienced trampers. Definitely a fine-weather trip only.

Follow Route 4 to Lake Peel then climb the cirque wall beyond the lake to the ridge crest, passing through bluffy outcrops of Devil River Volcanics, to the Mount Peel ridge. This is the highest point at which the old Northwest Nelson peneplain surface can be recognised and from here a fine view of the now rather dissected peneplain can be seen, looking towards Mount Arthur and the Karamea Fault across the Tableland. The view up and down the valley shows its smooth-sided glaciated form. Next descend the large scree slope formed of Tasman Formation siltstone debris and cross the saddle to continue along Peel Range. The route passes through dark siltstone containing scattered and poorly preserved graptolites and then passes into the Summit Limestone of lower Ordovician age. The limestone contains rare microfossils but is largely crystalline; originally, it probably contained macrofossils such as brachiopods and cephalopods. It forms a karrenfeld with sinkholes and caves and extends out onto Mount Mytton itself. A short detour to further along Peel Range brings us to the Anatoki Thrust and the boundary of the Central Belt. Across the thrust are strongly cleaved slate, siltstone and quartzose standstone of the Western Belt and the ridge crest soon becomes jagged and slow to traverse. From the summit of Mount Mytton descend through the karrenfeld to the ridge leading ESE towards Thorns Creek, then drop down to the Thorns Creek flat from which a marked track returns to the Cobb Valley bottom track (Route 1).

Fig 34. Cross-bedded calcareous grits, deposited on land above sea level in late Cambrian time. Photo: LL.

ROUTE 6—MOUNT PROSPECT AND MOUNT RANOLF (1 day)

A fairly strenuous trip on an unmarked route along the open tops—suitable for experienced trampers in fine weather only.

Follow Route 1 to Chaffey Stream, then climb the long gentle ridge on the north side of Chaffey Stream to the bush edge. Outcrops of Western Belt sandstone and cleaved siltstone are seen, and in some outcrops cleavage is so strongly developed that the rocks are termed slates, and original bedding is difficult to see. At the bush edge the rocks are mapped as Leslie Formation, and the route passes westwards still climbing, but passing down through this unit to the base where a massive quartzite bed is seen. The quartzite bed is one of several closely spaced quartzites which can be traced all the way from Mount Olympus down to Wangapeka Valley. Looking to the south, you will see three separate quartzite bands clearly (Fig. 35). The quartzite band is bordered to the west by a fault, and the route then passes through a thick succession of alternating grey to grey-green sandstone and siltstone (Roaring Lion Formation) which continues westwards to the Roaring Lion River. Mounts Prospect and Ranolf are formed from rocks of the Roaring Lion Formation. From Mount Ranolf be careful to take the eastern ridge and not the southern one leading to Kimbell Spur. The eastern ridge leads back through the rock sequence just traversed and soon the massive quartzite band is encountered again. Beyond lie the alternating quartz sandstone and black shale with abundant graptolites of the Leslie Formation. The graptolites are best seen in the scree slopes on the south side of the ridge and show clearly on the naturally weathered slabs. Descend to the Cobb Valley bottom via the ridge on the north side of Hannah Creek, and down the valley via Route 1.

Fig. 35 Quartzite beds in the base of the Leslie Formation form prominent ribs in the south wall of the valley of Chaffey Stream. These are clearly seen from Route 6. At this locality, there are three beds, but generally there are only two. Photo: RAC.

ROUTE 7—FENELLA—WAINGARO PEAK—MOUNT BENSON (1 day)

This route is done from Fenella Hut at the head of Cobb Valley. The route is not strenuous but is largely unmarked along the open tops and is suitable for experienced trampers.

Fenella Hut is sited near the low saddle between the Cobb and Burgoo Rivers, and it is well worthwhile visiting the saddle with its picturesque tarns from which one looks down into the impressive head of the Burgoo. The saddle marks an old glacial connection between the Cobb and Burgoo glacial systems. A poled route leads from Fenella Hut to the slopes of Waingaro Peak (known to geologists as Graptolite Hill), on the southern ridge of which abundant graptolites are found in black slates of the Leslie Formation. Following the ridge crest southwards the route passes along alternating quartzites and shale bands of the Leslie Formation which form small parallel strike ridges.

Graptolites can be found at several localities along the ridge as one moves SE but eventually we leave the Leslie Formation and pass into the strongly cleaved siltstone and sandstone of the Douglas Formation. From the ridge crest the rugged ridge lines formed of Western Belt rocks are seen to the north and west. As Mount Benson is approached we cross the Anatoki Thrust and here it is defined in a spectacular manner, by a "blade" of Summit Limestone that stands up prominently (Fig. 17), forming a feature known to geologists as "The Gendarme". The limestone is involved along the fault for much of its length in a zone of complex faulting and folding. The Mount Benson "massif" is formed of sandstone, grit and volcanics. On the far side of Mount Benson we reach the Cambrian limestone band which marked the end of the Route 3 traverse. We can descend to the valley floor here and return to Fenella Hut along the main track up the valley. The route passes back through the sequence just traversed but outcrops along the track are few until the climb from Cobb Hut up to Fenella.

ACKNOWLEDGEMENTS

This guide book is based on the work of several people who, for lack of space, cannot be acknowledged in the text. However, I acknowledge here the work of my colleague George Grindley who has contributed more than anyone else to our present understanding of the geology of Northwest Nelson.

I thank the following for their comments on the MS: Bruce Hayward, Miles Reay, Jane Forsyth, Graeme Stevens, George Grindley, Mike Johnston, Jane Soons, Ruth Darling, Jim Lowery, and Bob Dickson. Technical assistance from Wendy St George (photography), Stephan Spiekerman (draughting), Fay Tonks (typing) and Mary Cresswell (editing) is gratefully acknowledged. Bob Dickson (NZ Forest Service), Peter Riley (Lime and Marble Ltd) and John Simes (NZ Geological Survey) helped with various aspects of field work. Lands and Survey Dept gave permission to reproduce Fig. 30, and NZ Electricity Division, Fig 28. Other photographs were taken by D.L. Homer (DLH), L. Laronde (LL), Wendy St George (WStG), and the author (RAC).

FURTHER READING

For a general coverage of the early Paleozoic geology of New Zealand, the reader is referred to "The Geology of New Zealand" edited by R.P. Suggate, G.R. Stevens, and M.T. Te Punga and published by the Government Printer, Wellington. The articles listed below are available at specialist scientific libraries, such as at the universities and Geological Survey of the Department of Scientific and Industrial Research in Lower Hutt. However, for those who do not have access to these libraries the publications can be borrowed through a recognised library via the library interloan service.

Benson, W.N. 1956: Cambrian rocks and fossils in New Zealand. *Proceedings of the 20th International Geological Congress 2(2),* pp. 285-288. (First record of the discovery of Cambrian fossils in New Zealand.)

Clark, R.H., Vella, P. and Waterhouse, B.W. 1967: The Permian at Parapara Peak, Northwest Nelson. *New Zealand Journal of Geology and Geophysics* 10, pp. 232-246. (Describes the Permian rocks and fossils at Parapara Peak.)

Coleman, A.C. 1971: Geology of the middle Cobb area, Northwest Nelson. Thesis (B.Sc. Hons) held at University of Otago, Dunedin, New Zealand. (Describes the geology of part of the Central Belt sequence. Unfortunately not readily available.)

Cooper, R.A. 1979: Lower Paleozoic rocks of New Zealand. *Journal of the Royal Society of New Zealand* 9(1), pp. 29-84. (A general account of the Lower Paleozoic geology and fossils of New Zealand.)

Grindley, G.W. 1961: Sheet 13—Golden Bay (1st ed.). "Geological Map of New Zealand, 1:250 000". Department of Scientific and Industrial Research, Wellington, New Zealand. (Although a little out of date, this is still a useful map of the whole NW Nelson region.)

Grindley, G.W. 1980: Sheet 13—Cobb (1st ed.). "Geological Map of New Zealand, 1:63 360". Department of Scientific and Industrial Research, Wellington, New Zealand. (The most detailed geological map covering the Cobb Valley and Mount Arthur Tableland.)

Grindley, G.W. and Watters, W.A. 1965: Geology of Nelson. Chapter 3 in *National Resources Survey Part IV, Nelson Region.* (General outline of geology and economic mineral deposits of the region).

Hall, W.E., and Reed, J.J. 1945: Chrysotile asbestos deposits, Flora Survey District, Northwest Nelson. *New Zealand Journal of Science and Technology* 27B, pp. 177-184. (Describes the asbestos occurrence; somewhat dated.)

Henderson, R.A., and MacKinnon, D.I. 1981: New Cambrian inarticulate Brachiopoda from Australasia and the age of the Tasman Formation. *Alcheringa* 5, pp. 289-310. (Describes the inarticulate brachiopods from the Cambrian of Cobb Valley. Excellent illustrations.)

Hunter, H. 1977: Geology of the Cobb Intrusives, Takaka Valley, Northwest Nelson, New Zealand. *New Zealand Journal of Geology and Geophysics* 20, pp. 469-502. (Detailed description of the ultramafic rocks now called the Cobb Igneous Complex.)

Keble, R.A., and Benson, W.N. 1929: Ordovician Graptolites of Northwest Nelson. *Transactions of the New Zealand Institute* 59, pp. 840-863. (The first, and now rather outdated, account of the Cobb graptolites.)

Skwarko, S.K. 1962: Graptolites of Cobb River—Mount Arthur Area, Northwest of Nelson, New Zealand. *Transactions of the Royal Society of New Zealand* 1(15), pp. 215-247. (The most recent account of the Cobb graptolites.)

Soons, J.M. and Selby, M.J. 1982: *Landforms of New Zealand.* Longman Paul Ltd, Auckland. (Contains a good section by Johnston on Northwest Nelson.)

GLOSSARY

anticline: upfold in layered rocks; opposite of syncline
batholith: a huge intrusive body of granitic rock
bed, bedding: single layer of stratified rocks
brachiopods: marine animals with bivalve shells and common as fossils; similar in appearance to bivalved molluscs but quite different in basic structure
bryozoans: plant-like colonial marine animals; not related to graptolites or corals
calcite: the mineral calcium carbonate, $CaCo_3$; the main component of fossil shells and of limestone
carbonate: short for rock-forming minerals which contain carbonate (CO_3); e.g., calcium carbonate ($CaCO_3$)
chert: rock composed of amorphous silica
cirque: basin-shaped glacial valley, commonly with a tarn
cleaved, cleavage: planes of weakness developed in rock during folding and recrystallisation (i.e., not bedding)
conglomerate: sedimentary rock containing boulders and pebbles of older rocks
correlation: establishing different rocks as being of the same geological age
deformational fabric: pattern of fractures, cleavage, schistosity, &c., in rocks that have been deformed during faulting and folding
diamictite: conglomerate with fragments ranging from boulders to sand in a muddy matrix
fault: a fracture along which blocks of the Earth's crust have moved relative to each other
geological time scale: the sequence of geological time units; e.g., Paleozoic, Mesozoic, Cenozoic, &c. (See Fig. 5 and Table 1)
glacial period: cold climatic phase when glaciers and ice sheets grow
hanging valley: tributary glacial valley perched above the floor of a main glacial valley
igneous rocks: rocks formed by solidification from a molten or semimolten state
interglacial: the warm period between glacial periods
intrusive rocks: rocks formed when molten rock is injected or intruded into other rocks
iron pan: a hard layer beneath soil surface; formed by iron oxide leached from the upper layers
isotopes: elements which differ only in their number of neutrons; e.g., normal carbon, ^{12}C, has isotopes ^{13}C and ^{14}C
karrenfeld: type of landscape developed on limestone or marble, with caves, sinkholes &c.
laterite: hydroxides of iron and/or aluminium produced by strong weathering and leaching of soil
mafic: rocks and minerals rich in magnesium and iron
matrix (of rocks): the ground mass, or fine-grained component
metamorphism: change in the texture or composition of a rock; produced by heat and pressure at depth
molluscs: a division of the animal kingdom which contains pelecypods (e.g., mussels, cockels, &c., snails, octopuses, and chitons.
moraine: unsorted mixture of boulders and silt deposited by glaciers
orogeny: major period of mountain building, with folding and faulting of rocks and, commonly, emplacement of granite
peneplain: a land surface worn down by erosion to a nearly flat plain
peridotite: intrusive igneous rock rich in mafic minerals and lacking feldspar
polyp: single individual of a colonial animal, such as a coral
radioactive (element): an element that can change into another element by emission of charged particles from its nucleus
roche moutonnée: glacially smoothed knob or projection of bedrock

schist: metamorphic rock with mica minerals aligned
sedimentary rocks: rocks formed from sediments deposited by rivers, sea, wind or glaciers, &c.
sequence (rock sequence): the successive layers of sedimentary rock strata
serpentinite: rock composed of serpentine, a hydrated magnesium silicate
silled: having a sill or ridge across the outlet (of a cirque or basin); caused by glacial overdeepening
solifluction debris: rock waste formed in cold-climate regions and mantling the hillsides
strata: layered or bedded rocks
stratigraphic: referring to the sequence of rock strata
syncline: downfold, with a V- or U-shaped profile in layered rocks
tableland: a flat or undulating uplifted land area
thrust fault: fault along which a rock mass is pushed up and over an adjacent rockmass
topography: form and features of the surface of a region
ultramafic: rocks and minerals abnormally rich in magnesium and iron